AF799663

Topologies as Techniques for a Post-Critical Rhetoric

Lynda Walsh • Casey Boyle
Editors

Topologies as Techniques for a Post-Critical Rhetoric

palgrave
macmillan

Editors
Lynda Walsh
Department of English
University of Nevada
Reno, Nevada, USA

Casey Boyle
Department of Rhetoric and Writing
University of Texas at Austin
Austin, Texas, USA

ISBN 978-3-319-51267-9 ISBN 978-3-319-51268-6 (eBook)
DOI 10.1007/978-3-319-51268-6

Library of Congress Control Number: 2017932645

© The Editor(s) (if applicable) and The Author(s) 2017
This work is subject to copyright. All rights are solely and exclusively licensed by the Publisher, whether the whole or part of the material is concerned, specifically the rights of translation, reprinting, reuse of illustrations, recitation, broadcasting, reproduction on microfilms or in any other physical way, and transmission or information storage and retrieval, electronic adaptation, computer software, or by similar or dissimilar methodology now known or hereafter developed.
The use of general descriptive names, registered names, trademarks, service marks, etc. in this publication does not imply, even in the absence of a specific statement, that such names are exempt from the relevant protective laws and regulations and therefore free for general use.
The publisher, the authors and the editors are safe to assume that the advice and information in this book are believed to be true and accurate at the date of publication. Neither the publisher nor the authors or the editors give a warranty, express or implied, with respect to the material contained herein or for any errors or omissions that may have been made. The publisher remains neutral with regard to jurisdictional claims in published maps and institutional affiliations.

Cover image © Thomas Koidhis/Getty Images

Printed on acid-free paper

This Palgrave Macmillan imprint is published by Springer Nature
The registered company is Springer International Publishing AG
The registered company address is: Gewerbestrasse 11, 6330 Cham, Switzerland

Contents

Contributor Information, Topologies as Techniques for a Post-Critical Rhetoric

Co-editors

Casey Boyle is an assistant professor in the Department of Rhetoric and Writing at the University of Texas, Austin, where he researches and teaches digital rhetoric and media theory. His work has appeared in *Computers and Composition*, *Technical Communication Quarterly*, *College English* as well as essay collections *Rhetoric and the Digital Humanities* and *Thinking with Bruno Latour in Rhetoric and Composition*. Right now, Casey is completing his first book, *Rhetoric as a Posthuman Practice*, that explores the role of practice and ethics in digital rhetoric.

Lynda Walsh is Associate Professor of English at the University of Nevada, Reno. Her research field is the rhetoric of science, specifically the interface between science and the public. She has published two monographs on these subjects using topological methods: *Sins Against Science: The Scientific Hoaxes of Poe, Twain, and Others* (2006); and *Scientists as Prophets: A Rhetorical Genealogy* (2013). She has also published articles on scientific controversies, visual rhetoric, environmental rhetoric, and non-Western rhetoric.

Contributors

L. Gregory Appelbaum is an assistant professor in the Department of Psychiatry and Behavioral Sciences at Duke University School of Medicine, Durham. Appelbaum's research interests primarily concern the brain mechanisms underlying visual cognition, how these capabilities differ among individuals, and how they can be improved through behavioral, neurofeedback, and neuromodulation interventions. Within the field of cognitive neuroscience, his research has addressed visual perception, sensorimotor function, executive function, decision-making, and learning/ expertise.

Elizabeth Beam is a graduate student in the Stanford School of Medicine. She is interested in neuroimaging both as a subject for meta-analysis and as a tool for elucidating mechanisms of mental illness. Her current research at Stanford Medical School applies machine-learning techniques to functional connectivity networks in order to identify and diagnose novel subtypes of post-traumatic stress disorder.

Adam W. Cody is a graduate assistant in Communication Arts and Sciences at the Pennsylvania State University, State College, where he studies rhetorical argumentation and the discursive habits of democratic pluralism. His current projects focus on reading ancient theories of reasoning and language in conversation with their particular political and cultural environments.

Jake Cowan is a doctoral candidate in the English Department at the University of Texas, Austin, with a specialization in rhetoric, writing, and digital literacies. His research is situated at the intersection of post-structural Lacanian psychoanalysis and rhetorical criticism, with a specific interest in construction of subjectivity through media ecologies.

Rosa A. Eberly is an associate professor in Communication Arts and Sciences and English at The Pennsylvania State University, State College. A free-range rhetorician, Eberly was formerly associate professor and director of the writing center at The University of Texas, Austin. Author of *Citizen Critics: Literary Public Spheres,* a major work in topological analysis, and co-author of *The Elements of Reasoning*, she is also co-editor of *The Sage Handbook of Rhetorical Studies* and *A Laboratory for Public Scholarship and Democracy.*

William Hart-Davidson is an associate professor in the Department of Writing, Rhetoric, and American Cultures and Associate Dean of Graduate Education in the College of Arts & Letters at Michigan State University, East Lansing. He is co-editor, with Jim Ridolfo, of *Rhetoric & the Digital Humanities* (2015). His work on computational rhetoric has appeared recently in *The Journal of Writing Research*, *Enculturation: Journal of Rhetoric, Writing, and Culture*, and *dhcommons journal*

Molly Hartzog is an assistant professor at Frostburg State University. During her PhD work at North Carolina State University, she was an NSF-IGERT fellow in NCSU's Genetic Engineering and Society program where she works with an interdisciplinary team of students on the social, ethical, and economic considerations for using genetic pest management for disease control. Her dissertation explored the use of genome sequence databases as tools for rhetorical invention in genetic engineering.

Scott A. Huettel is the Jerry G. and Patricia Crawford Hubbard Professor in the Department of Psychology and Neuroscience at Duke University, Durham. Huettel's research uses a combination of behavioral, genetic, physiological, and neuroscience techniques to discover the neural mechanisms that underlie higher cognition, with a focus on economic and social decision-making. He is an author of more than 100 scientific publications, including articles in *Science, Nature Neuroscience, Nature Reviews Neuroscience, Neuron, Psychological Science*, and other top journals in several fields.

Jordynn Jack is a professor in the Department of English and Comparative Literature at the University of North Carolina, Chapel Hill. Jack is author of *Science on the Home Front: American Women Scientists in World War II* (2009), *Autism and Gender: From Refrigerator Mothers to Computer Geeks* (2014), and editor of *Neurorhetorics* (2012). Her current research focuses on "neurorhetoric," or the problem of persuasion in neuroscience, and on interdisciplinary health humanities projects investigating the health effects of narrative writing among people with diabetes and aging populations.

James Moody is the Robert O. Keohane Professor of Sociology at Duke University, Durham. Moody has published extensively in the field of social networks, methods, and social theory. He has used network models to help understand school racial segregation, adolescent health, disease spread, economic development, and the development of scientific disciplines. Moody's work is funded by the National Science Foundation, the National Institutes of Health and the Robert Wood Johnson Foundation and has appeared in top social science, health, and medical journals.

Ryan Omizo is an assistant professor in the Department of Writing and Rhetoric at the University of Rhode Island, Kingston. His research interests include computational rhetoric, Asian-American rhetoric, and the digital humanities. His work has appeared in *The Journal of Writing Research, Enculturation: Journal of Rhetoric, Writing, and Culture, dhcommons,* and ACM SIGDOC.

Lawrence J. Prelli is Professor of Communication at the University of New Hampshire, Durham. Prelli's work focuses on the rhetoric of science, particularly rhetorics of display in public debates. He revived topological approaches to technoscience with his seminal text *The Rhetoric of Science* (1989), and his articles have appeared in all of the top communication journals. His recent edited collection, *Rhetorics of Display* (2006) is a central work in the field of visual rhetoric.

Sara Rubinelli is Assistant Professor in Health in Health Sciences with a focus in health communication at the Department of Health Sciences and Health Policy of the University of Lucerne (Switzerland). She holds a degree in Classics and Philosophy from the Catholic University of Milan and a PhD from the University of Leeds in the areas of argumentation theory, persuasion and rhetoric. Since September 2009 she is also Scientific Coordinator of the Human Functioning Unit at Swiss Paraplegic Research (Switzerland) and leads there the Person-Centered Healthcare Group. Since September 2016 she is President-elect of the European Association for Communication in Healthcare (EACH) and holds the scientific direction of the Observatory for Research and Practice of Medical Humanities of the Fondazione Sasso Corbaro (Switzerland).

Christa Teston is Assistant Professor of English at Ohio State University, Columbus. Teston studies backstage evidential construction and decision-making in medical and scientific contexts, and has published her research in various venues—including *Written Communication, the Journal of Medical Humanities, Technical Communication Quarterly,* and *Rhetoric Society Quarterly* (forthcoming). Christa is also a co-editor of the *New Directions in Rhetoric and Materiality* book series (The Ohio State University Press).

Kenneth C. Walker is Assistant Professor of Rhetoric and Composition, University of Texas, San Antonio. Walker teaches and researches in rhetoric of science and technology, technical and professional writing, environmental rhetoric, and digital media. His work is forthcoming or published in *Technical Communication Quarterly, Journal of Business and Technical Communication, Rhetoric Review, POROI,* and *Environmental Humanities.*

List of Figures

List of Tables

CHAPTER 1

From Intervention to Invention: Introducing Topological Techniques

Lynda Walsh and Casey Boyle

> Whereof one cannot speak, thereof one must remain silent.
>
> —Ludwig Wittgenstein (*Tractatus* 7)

Climate change, gender and identity, genetic engineering, globalization, homelessness, and automation and agency. Each of these discourses is consequential. Each is wickedly complex, resistant to traditional critical analysis, and stubbornly persistent. And importantly, each has habitual contours that amplify certain voices and arguments while muting others.

The epigraph hints at these contours, at habits of discourse so powerful and so ingrained that they take on the force of natural law. It also hints at the critical reflex that has produced little change in the wicked discourses listed above: first, dichotomize the discourse into two opposing terms (e.g., speech/silence); then, either synthesize the dichotomy via a transcendent term or show that one term is entailed by the other.

L. Walsh (✉)
Department of English, University of Nevada, Reno, NV, USA

C. Boyle
Department of Rhetoric and Writing, University of Texas at Austin, Austin, TX, USA

© The Author(s) 2017

L. Walsh, C. Boyle (eds.), *Topologies as Techniques for a Post-Critical Rhetoric*, DOI 10.1007/978-3-319-51268-6_1

The authors collected in this volume are all deeply interested in wicked discourses and deeply dissatisfied with the results of the critical reflex. They are also rhetoricians, which means they share a goal of working toward justice in public deliberation. Accordingly, these are the questions addressed in this volume: Given our current entanglement in global relations mediated through complex logics of technoscience, how do we solve problems together? How can we include and incorporate that which has been excluded and objectified? How can we reshape our present toward equitable futures? In short, *how do we move beyond* intervention *to* invention *in the wicked discourses that entangle us in our common lives?*

These questions prompt us to recognize that wicked discourses require new techniques that can work *through* (rather than around or above) the modern dichotomies of nature/culture, human/nonhuman, and text/context to create competent, inclusive accounts of dire political dynamics. *We propose that the practice of topology offers those techniques.* Topology combines the classical rhetorical strategy of *topos* with nonlinear logic to yield a new model of discourse that is (a) transductive in its working across concrete examples to illuminate structure,[1] (b) responsive to the contingencies of the discourse situation, and (c) generative of power dynamics that help shape that situation.

This book is an attempt to engage the possibilities of topologies as techniques for rhetorical invention. In the remainder of this introduction, we first review the current dissatisfaction with critique as a mode of humanist engagement with wicked discourses. We then provide a basic grounding in the ancient, generative notion of *topos* from its roots in military strategy and craftsmanship to its development as a regimen for fostering readiness to act in the face of uncertainty. Next, the movement from *topos* to topology is traced, and an orientation to the genealogy of topological theory in the humanities does double duty as an orientation to the chapters in this volume. Each contributor engages an aspect of rhetorical studies to explore, examine, and elaborate the possibilities of topological engagement; and, each treats a concrete case in order to display their methods, for the purposes of adoption and adaptation. We conclude the introduction with a selection of alternate topologies for the volume.

A Crisis in Critique

How can we as critics learn to compose instead of deconstruct? This has been the primary question confronting humanists in the decade or so since Donna Haraway, Bruno Latour, and others sounded the alarm on both a pivotal moment in and an unwelcome irony to the critical tradition.[2] The irony was that, in trying to transcend via dialectics the subject/object dichotomy and all its troublesome offspring (nature/culture, male/female, white/black, self/other, colonizer/colonized, etc.), post-modern critics merely reified the dichotomy by packing its problematics into its first term, subjectivity.[3] The pivotal moment was the post-human turn, in which subject and object became so knotted and doubled that they would no longer yield to the blade of the dialectic. New-materialist and post-critical scholars called for new methods that didn't depend on "breaking through" or "rising above" to help humans and nonhumans live justly together in uncertainty.

The solution, or at least one of them, lay hidden within the critical reflex itself. It is visible in the Wittgenstein epigraph above, in the words "*where*of" and "*there*of," which trace a spatial continuity across the apparent discontinuity posed by the dialectic. The spatial turn in criticism had its groundwork laid nearly a century ago in the work of Ernst Cassirer[4]; was given a critical vocabulary by Michel Foucault (via the disciplinary functions of the panopticon) and Jacques Lacan (via topology and psychoanalysis)[5]; and finally came to prominence in Henri Lefebvre's and Edward Soja's theories of urban spaces as social productions.[6] The warping, folding dynamics of spatiality promised new ways to follow the knots of a wicked discourse and tease out change without having to resort to the scissors of the dialectic.

By the early 1980s, the spatial metaphor had become epidemic in the humanities: science-studies scholars, philosophers, design theorists, geographers, economists, and psychoanalytic critics all eagerly explored this new dimensionality for criticism—flat and associative rather than hierarchical and reductive.[7] The chief motivation was to account for all the actors in wicked discourses—ideas, states, people, technologies, and "nature," to see them interact in analysis, and to make room for change. Sweeping, reductive theories of causation were quickly replaced with "myopic," hybrid narratives that composed networks of alliance and resistance across real and figurative spaces.[8] With these new methods, post-critical scholars hoped to reveal how power dynamics were constituted via a traceable account that would permit not only their deconstruction but also their recomposition along alternate lines.

From Intervention to Invention

Brian Rotman begins his examination of the possible correspondence between mathematical and cultural topology with a question and two responses. "What is a topological space? Two responses: one, palpable and familiar examples, the other abstract and alien."[9] In other words, topology takes what is familiar and transforms it until it is alien but still of a piece with itself. In mathematical topology, inquiry stretches, twists, folds, and bends a form to its limits—cuts and ruptures being the only disallowed transformations. In humanist applications, a topological approach traces the contours of a discourse and may fold it into a new configuration. Thus, change may be invented without resorting to the ruptures of critical intervention.

From the Greek for "place," "space," or "stance," the notion of *topos* as strategic position has always proven fertile ground for invention (see Glossary for more on the *topoi*). Aristotle himself invoked spatiality when he imagined arguing from *topoi* as "selecting" among elements "inherent to" or "intimately connected with" a given discourse (*Rhetoric* 2.22). These elements included generic schema like compare/contrast and cause/effect; shared beliefs, norms, and values; and, foundational principles of the discourse at hand, whether physics or tax law. This spatial orientation extended beyond the *Rhetoric* to the *Physics*, informing Aristotle's discussion of the mechanics of change. *Topos* was the ground against which change could be recognized, as "place could not be either the matter or the form of the thing contained, but must be different" (210b 27–28). Here was a central irony: while different from matter and form, place was nonetheless fundamental to its definition. Luce Irigaray articulated this irony as follows: "[p]lace would in some way be the 'nature' of matter and form, the habitat in which both wed without ceasing, and in their extension. To infinity."[10] By serving as the common ground against which change could be recognized and traced, place yielded the continuous thread from which the twin myths of matter and form could be spun. Likewise, in rhetoric, *topos* yielded the common ground that enabled the myth of community.

It was not such a great leap from this ancient doctrine of continuity-in-change to humanist topologies. By welding Aristotle's ironic notion of *topos* to mathematical topologies that reimagined "form" as a function of elastic relations among nodes, post-critical scholars fashioned a way to model discourse without dichotomizing it and/or reducing it to transcendent causes such as "society" or "capitalism." Promising and exciting, as we can sense in the buzz generated by Sloterdijk's "bubbles" and Bennett's "vibrations,"[11] these systems nonetheless present a problem: they have forgotten that the roots of topology lie in rhetoric, not philosophy. This category mistake matters tremendously because without their rhetorical rooting, topologies lose their heuristic qualities and thus their ability to compose—not just deconstruct—political dynamics.

Topologies achieve their heuristic effect via an ancient rhetorical tension between *topos* and *kairos*. *Kairos* means an opportune moment; it has both a military derivation, as in a propitious opening in an opponent's defenses, and a craft derivation, as in the moment in weaving when the warp strands open on the loom to allow the shuttle to carry the weft strand *through* (see Glossary for more on *kairos*). It is easy to construe *kairos* here as a rupture—and certainly some theorists have[12]—but our understanding of *kairos* is informed by Debra Hawhee's examination of *it* as an "emergence" through "relational specificity."[13] In other words, we understand *kairos* not as a break in habitual relations but as an emergent fold or wrinkle in them that opens a space for reflection and revision.

In politics, *kairoi* are emergent opportunities to modify communal life that cannot be precisely predicted, perfectly controlled, or exactly replicated by any art or philosophy. Instead, they demand a "response-ability," a disciplined readiness. This readiness, according to the classical rhetorical tradition is, in part, provided by training in the *topoi*; that is, if actors habituate themselves to the values, norms, and beliefs "inherent to a subject" as well as the twists and turns arguments tend to take, they position themselves to profit from emergent *kairoi*. That essential tension between *kairos* and *topos*, between chance and art, made classical systems of *topoi* capable of producing—not merely interpreting—political effects. The tension became attenuated, however, in Continental treatments of Aristotle's work, in which *topoi* gradually devolved into and were dismissed as "commonplaces" that betrayed a particularly bourgeois failure of imagination. This reduction of *topoi* to static formulae[14] carried into and infected even the most dynamic concepts of topology, such as Heidegger's, which, while it purports to describe the situational contours of being that make us emerge for ourselves and for others as beings, still fails to understand that emergence as a fundamentally political act.[15] The political vectors of topology must be restored in order for them to serve as truly post-critical techniques. Otherwise, they will simply collapse back into the critical tradition as fancy variations on a post-structuralist theme.

As we look to reconnect topology to rhetoric, our project resonates with a host of others outside our field. In particular, we find affinity in John Allen's definition of topological power as one in which "[t]he 'changing same' of power represents one side of the topological equation where a relationship of power is often reproduced through different registers"[16] but which also affords the supposed subjects of that power to be differently empowered. We also are in conversation with the "*becoming topological* of culture"[17] as expressed by Celia Lury, Luciana Parisi, and Tiziana Terranova and look to align our project with its aims. Further, we are coming to recognize the extent to which topological constructs bolster our network and financial complexities.[18] What these diverse sources show is that topologies are becoming a wider cultural technique than may be reduced to any one discipline. We return to topologies' roots in rhetorical practice in order to help our colleagues in all fields better invent alternatives in wicked discourses.

The rhetoricians collected in this volume work across the full range of the topological tradition. To each strand they restore the essential tension between *topos* and *kairos*, giving it the force to generate possibilities for a more just life together in our post-human moment.

Shaping Topologies

In this section, we trace the volume contents along with the development of topology in the humanities. Naturally, there are other topologies the reader might wish to trace through the volume; we suggest a few of these at the conclusion of the introduction. For more on the classical terminology, see Glossary that precedes the Index to this volume.

Classical Topology

Deriving topology from Aristotle's *topoi*, both from the *Rhetoric* and the *Physics*, is not automatic. While the germ of topology is certainly present in his work—particularly in the special topics (*idia*) or foundational principles of a given disciplinary discourse—what work remains from Aristotle never fully realizes the spatial promises of his system. Sara Rubinelli argues that a proto-topology can be recovered from Aristotle as a model of reality that audiences agree to accept for the purposes of engaging in argumentation. Rubinelli establishes this concept through an exacting examination of Aristotle's topical system in the *Topics* and *Rhetoric*.

Along related lines, Cody and Eberly derive topology from classical texts by putting the *tekmēria*—pre-existing boundary markers—into tension with the strategic stances articulated in the *topoi*. By analyzing Aristotle against Isocrates and Alcidamas, Cody and Eberly convincingly demonstrate that *tekmēria* provide necessary bounds on topology, and also that these bounds will shift with time and practice. Note that Cody and Eberly disagree with Rubinelli regarding the relationship of the universal and special topics: where Rubinelli adheres to the traditional hylomorphic reading—wherein the common or universal topics supply the form of an argument while the special topics provide its content—Cody and Eberly envision a more holistic topical doctrine. Such differences of interpretation are inevitable when scholars work with a 2,500-year-old theory.

Like Rubinelli, the Italian Marxist philosopher Paolo Virno saw a distinction between Aristotle's common and special *topoi*. He worked from it to argue that we are missing robust "special places" from which we might base collective action in times of economic change. Responding to Virno, Casey Boyle suggests that the notion of commonplace, in the context of sharing economies, undermines the possibility of sustained labor collectives. Using the recent examples of sharing economy labor disputes including *Uber*, a ride sharing company and app, Boyle shows how topological assemblages might help reshape labor efforts. That is, while new and emerging corporations are drawing upon the shifting and placeless nature of new media, labor too must become reshaped in an era of emerging *nonemployment*.

Finally, Molly Hartzog presents an unconventional, but compelling, view of the Aristotelian landscape. She posits that the universal *topoi* may operate at a meta level on the special *topoi* of a discourse to define acceptable contributions to it. She first identifies the technical ontologies of a mosquito-disease-vector database as topologies built from special epidemiological *topoi*. Then, she applies Aristotle's universal *topoi* as a lens on those ontologies to show they privilege disciplinary consensus over the discovery of new vectors in mosquito-borne disease. This topological revelation of technical habits of thinking promises opportunities for invention in cases where group-think may fence out creative solutions to problems.

Mathematical Topology

Mathematicians pushed Aristotle's spatial reasoning into a different technical arena when they developed topology as a means to solve complex geometric problems. While a few stray references to topology surface in seventeenth-century botany and geography—in somewhat of a tangle with "topography," the first extended scientific engagement with topology appears in mathematics in the late nineteenth to early twentieth centuries. Johann Benedict Listing coined the term "topologie" in 1847 to describe ways of treating mathematical functions so that their characteristics were preserved against scalar transformations. However, most accounts of mathematical topology begin earlier, with Leonhard Euler's famous "Seven Bridges of Königsberg" problem (1750)—in which he reduced the city to a network of nodes to demonstrate that one could not make a circuit of the town while only crossing each of its seven bridges once.

The classical example of a mathematical topology lies in the joke that a topologist eats his doughnut and finds his coffee mug missing because a mug and a doughnut are two permissible deformations of the same form—a torus. Mathematical topologies redefine form not as a recognizable, static shape but as an elastic assemblage of nodes and "edges," or connections among nodes. In this volume, William Hart-Davidson and Ryan Omizo apply mathematical topologies quite literally to the study of rhetoric—using word association frequencies to create a second-order graph of a text as a series of nodes (words) and edges (associations). Returns to certain nodes over time—that is, "loops" or "walks"—constitute *topoi*. The networked *topoi* of a discourse instantiate its topologies. Stable topologies constitute genres, and computational tracking of mutations in these topologies over time allows us to observe the process of genre formation at a level of concreteness that may enable us to predict and shape it in limited ways.

Jack et al. also employ computational topologies to explore an interesting conflation of "topology" and "topography" in neuroscience. With an ingeniously reflexive method based on Social Network Analysis (SNA), they show how neuroscientists assure themselves of robust functional connections in the brain by tracing topologies of citations across neuroscience literature. Jack and colleagues argue that these hybrid linguistic/neurological topologies both constrain how neuroscientists conceptualize brain functions and serve to channel work and attention in future research.

Lacanian Topology

Several twentieth-century humanists made the connection between mathematical and verbal topologies; the most famous of these was the psychoanalytic critic Jacques Lacan. Working from topologies such as the Möbius strip and the Borromean knot, Lacan achieved the insight that the signifier and the signified, and therefore interiority and externality, articulated a differential space of nodes and edges that could be deformed but not separated. Lacan's students Luce Irigaray and Félix Guattari, among others, elaborated on his spatial theories.[19] In this volume, Jake Cowan engages this tradition to argue that Lacan was in fact turning toward rhetoric in his topological phase. In a close reading of Lacan's late seminars, Cowan suggests that Lacanian topologies provide a rubric for writing that can articulate contradictory identities and others without reducing them.

Constructivist Topologies

At around the same time that Lacan was pacing the streets of Paris worrying knots of string, Michel Foucault was elaborating his own spatial theory of power dynamics based on Greek philosophy. Although Foucault never used the term topology, his "grid of intelligibility" is nevertheless a topical approach to understanding how regimes of power impose themselves on discourse; they generate matrices of controlling *topoi*—like male/female for sexuality—that determine which ideas may be discussed and how.[20]

Multiple scholars have found these notions productive and have elaborated topologies based upon them. The spatial turn in sociology worked from Foucault and from mathematical topology to posit social structure as a network of nodes and edges.[21] Tracing the development of these networks over time revealed the accretion of power in certain nodes. However, these constructivist topologies lacked the internal mechanics of *topoi*, which can reconfigure to invent alternatives to ossified power relations.

Three contributors to this volume take up the challenge to put rhetorical teeth into constructivist frameworks. Kenneth C. Walker activates Actor Network Theory to track the metaphor of James Hansen's "Loaded Climate Dice" across activist and skeptical media. The resulting topology reveals that the metaphor was not well adapted to the contours of skeptical discourse, thus reifying rather than dissolving political boundaries in climate change debates. The case contains lessons for scholars and practitioners hoping to deploy metaphors as "common places" among diverse constituencies.

Lynda Walsh and Lawrence J. Prelli make a novel application of classical *topoi* to technical graphics in order to demonstrate how the Foucaultian synoptic or "view from nowhere" was constructed topologically in early US ecology. They implicate synoptic topologies in the move to biopolitics and suggest ways in which activists might recognize and subvert related moves via the careful application of resistant *topoi*.

Finally, Christa Teston treats oncologists' diagnostic practices as a negotiation between chance and art, a "tacking back and forth" between known and unknown disease experiences to compose a topology of treatment. Teston notes that while this topology can seriously limit the ways oncologists compose their patients' bodies and disease experiences, it also exhibits an "enthymatic elasticity" that holds some promise of seeing hegemonic medical topologies shift and even break apart over time.

Imagining Alternatives

While the contributors to this volume apply topology in a range of ways to *kairoi* ranging from labor to genre studies, nevertheless, all of the collected approaches perform the central commitments of rhetorical topologies:

- Rhetorical topologies work across concrete verbal and visual instances, not with the goal of inducing generalizations, but with the goal of transducing the unique contours of a discourse at a specific time and place. By refusing to reduce complex *kairoi* to their causes, topologies adhere to post-critical ideals. And by remaining stubbornly concrete and traceable, topologies maintain rhetoric's traditional moorings in case study and close reading, anchors that have for centuries made our discipline an indispensable dialectical partner to disciplines such as philosophy, history, sociology, and literary criticism.
- Rhetorical topologies maintain constant tension with *kairos*. They perform the fundamental ambivalence between chance and art that lies at the very root of Greek topical systems. At the same time, topologies reveal that *kairoi* are not rips, tears, or ruptures in habitual practice, as they are so often believed, but are rather folds in that practice that frame fleeting opportunities for structural change. While topologies are durable, reflecting their communities' reflexes for dealing with uncertainties, they also shift over time and practice.
- Rhetorical topologies generate political dynamics. They do so because *topoi* are Janus-faced, serving not only as hermeneutics but also as heuristics. Here lies their strongest contribution to post-critical theory: by giving an explicit account of the generation of political dynamics in *kairos*, topologies provide a guide to inventing alternative dynamics: whether by articulating the actors and arguments hiding in the unarticulated gaps and margins of topologies;

by deploying counter-*topoi* to deflect dominant *topoi* in new directions; or, by shifting from intervention to invention, that is, from the transcendent and synthetic mode of the dialectic to the iterative and associative mode of the topologic.

Finally, as one of the goals of this volume is to call attention to the hidden topologies of seemingly objective discourses, it seems only right to provide some alternative topologies by which this volume could be read and understood.

Archive
All of our rhetoricians treat concrete texts and cases, and as Foucault has argued, archives themselves are assembled according to a topology that inscribes societal value and order. Following an archival topology we would read:

Classical:	Eberly and Cody, Rubinelli
Disciplinary (STEM):	Teston, Walsh and Prelli, Hartzog, Jack et al.
Disciplinary (Humanist):	Cowan, Hart-Davidson
Mass media:	Walker
New media:	Boyle

Method
As all chapters are case based, there is a degree of empiricism to each. However, some scholars shift more toward qualitative methods in assembling their topologies, and some more quantitative.

Quantitative:	Hart-Davidson, Jack et al., Walker, Hartzog
Qualitative:	Boyle, Cowan, Walsh and Prelli, Rubinelli, Cody and Eberly, Teston

Construction Site
Most of our scholars are aiming to reinvent technical, disciplinary discourses, an observation which does not rule out an application to another sphere of discourse later on.

Technical:	Hart-Davidson, Jack et al., Teston, Hartzog, Rubinelli, Cody and Eberly
Private:	Cowan
Public:	Walsh and Prelli, Walker, Boyle

Scholarly Genealogy

This volume collects veteran scholars who participated in the topical revival in rhetoric and composition, with all its baggage and promise, alongside emerging scholars who are at ease thinking radically about *topoi* and topologies. To read from the critical past toward the post-critical future:

Walsh and Prelli
Hart-Davidson
Cody and Eberly
Rubinelli
Jack et al.
Boyle
Teston
Walker
Hartzog
Cowan

Notes

1. Transductive methods trace how a signal—be it an image, phrase, or discourse—continues across boundaries and through translations. Transduction takes into account how what has come before lays the ground for what comes after.

   The classical example of transduction is found in the recording and playback of speech, wherein a voice resounds from vocal cord vibrations, traverses the air through sound waves, manipulates a microphone's diaphragm, is translated into electrical signals, and is translated again into inscriptions (optical, magnetic, etc.) from which the whole process can be reversed to make that voice resound anew via the drivers of speakers, then travel through the air, cause vibrations of the tympanic membranes of the ear, be translated into electrical signals in the nervous system, and so on.

   Transduction emphasizes *continuity* rather than *identity*, which requires the reducing out or discounting of transformation. Thus, transductive approaches are responsive to contingencies and constraints of any given discourse situation. They also render an account of the power dynamics of that situation, as they reveal the structuring structure (to lean on Pierre Bourdieu) that prepares each movement from the prior one. In its deployment through and across multiple concrete situations, the logics underpinning topology are rightly characterized as transductive. As to the specific tactics and techniques of transductive approaches to topological analysis of discourse, those are what we pursue in this project. Cf. Rob Kitchin and Martin Dodge, *Code/Space: Software and Everyday Life.* (Cambridge, Mass.: The MIT Press, 2014); Pierre Bourdieu, *Outline of a Theory of Practice.* Trans. by Richard Nice. (New York: Cambridge University Press, 1977).

2. Donna Haraway, "Situated Knowledges: The Science Question in Feminism and the Privilege of Partial Perspective," *Feminist Studies* 14, no. 3 (1988); D.J. Haraway, *Modest–Witness@Second–Millennium. Femaleman–Meets–Oncomouse: Feminism and Technoscience* (Routledge, 1997); Gregory L. Ulmer, "The Art Object of Post-Criticism" in *The Anti-Aesthetic: Essays on Postmodern Culture*, ed. Hal Foster, ed., New York: The New Press, 1998, 95. Bruno Latour, "Steps toward the Writing of a Compositionist Manifesto," *New Literary History* 41, no. 471–490 (2010); "Why Has Critique Run out of Steam? From Matters of Fact to Matters of Concern," *Critical inquiry* 30, no. 2 (2004); John Muckelbauer. *Future of Invention, The: Rhetoric, Postmodernism, and the Problem of Change* (SUNY Press, 2009).
3. S Scott Graham, *The Politics of Pain Medicine: A Rhetorical-Ontological Inquiry* (University of Chicago Press, 2015).
4. Ernst Cassirer, "Mythic, Aesthetic and Theoretical Space," *Man and World* 2, no. 1 (1969).
5. M. Foucault, *Discipline and Punish: The Birth of the Prison* (Vintage Books, 1977), 26; *The History of Sexuality, Volume 1: An Introduction*, trans. Robert Hurley (New York: Random House, 1978), 93; J. Lacan and J.A. Miller, *The Four Fundamental Concepts of Psycho-Analysis* (W.W. Norton, 1998).
6. Henri Lefebvre, *The Production of Space*, vol. 142 (Oxford: Blackwell, 1991); Edward W Soja, *Thirdspace: Expanding the Geographical Imagination* (Oxford: Blackwell, 1996).
7. See in addition to the sources listed above for geography, Mike Davis, *City of Quartz: Excavating the Future in Los Angeles (New Edition)* (New York: Verso Books, 2006); for literature Georges Perec, "Species of Space and Other Pieces (Trans. J. Sturrock)," (Harmondsworth, UK: Penguin, 1997).; for design studies G. Teyssot, *A Topology of Everyday Constellations*

(MIT Press, 2013).; for philosophy P. Sloterdijk and W. Hoban, *Bubbles: Microspherology* (Semiotext(e), 2011).

8. Cf. Gilles Deleuze and Félix Guattari, *A Thousand Plateaus: Capitalism and Schizophrenia* (Bloomsbury Publishing, 1988); B. Latour, *Reassembling the Social: An Introduction to Actor-Network-Theory* (OUP Oxford, 2007).
9. Brian Rotman, "Topology, algebra, diagrams," *Theory, Culture & Society* 29, no. 4–5 (2012): 247–260.
10. L. Irigaray, *An Ethics of Sexual Difference* (Ithaca, NY: Cornell University Press, 1993), 37.
11. See note 5 above and Jane Bennett, *Vibrant Matter: A Political Ecology of Things* (Durham, NC: Duke University Press, 2009).
12. Lloyd F. Bitzer, "The Rhetorical Situation," *Philosophy & Rhetoric* 1, no. 1 (1968):1–14; Richard E. Vatz, "The Myth of the Rhetorical Situation." *Philosophy & Rhetoric* 6, no. 3 (1973):154–161.
13. Debra Hawhee, "Kairotic Encounters." In *Perspectives on Rhetorical Invention*, edited by Janet Atwill and Janice Lauer. Knoxville: U of Tennessee P, 2002, 32.
14. For current critiques about *topoi* as static and repetitive, see Thomas. *Ambient Rhetoric: The Attunements of Rhetorical Being*. University of Pittsburgh Press, 2013; Jeff Rice. *Digital Detroit: Rhetoric and Space in the Age of the Network*. SIU Press, 2012.
15. For Heidegger's derivation of topology from Aristotle, see Jeff Malpas, *Heidegger's Topology: Being, Place, World* (Cambridge, MA: MIT Press, 2008), 13–14; Otto Pöggeler, "Heidegger's Restricted Conception of Rhetoric," in *Heidegger and Rhetoric*, ed. Daniel M. Gross and Ansgar Kemmann (Albany: State University of New York Press, 2005); for Heidegger's reading of Aristotle's *Rhetoric* as phenomenological rather than political, see chapter 1 of Walter A Brogan, *Heidegger and Aristotle: The Twofoldness of Being* (SUNY Press, 2012).
16. John Allen. *Topologies of Power: Beyond Territory and Networks*. Abingdon, Oxon; New York, (Routledge, 2016): 4.
17. Celia Lury, Luciana Parisi, and Tiziana Terranova. 2012. "Introduction: The Becoming Topological of Culture." *Theory, Culture & Society* 29 (4–5): 3–35.
18. See, Alexander R. Galloway *Protocol: How Control Exists after Decentralization*. Cambridge, Mass.: MIT Press, 2004; Hannah K. Lee June, 2016. "Can Topology Prevent Another Financial Crash?—Issue 37: Currents." *Nautilus*. Accessed October 31. http://nautil.us/issue/37/currents/can-topology-prevent-another-financial-crash.
    Ford, *My Life and Work*, 267.
19. See note 6 above and L. Irigaray, *An Ethics of Sexual Difference* (Ithaca, NY: Cornell University Press, 1993), 34–56.

20. Michel Foucault, "The Order of Things," *New York: Vintage* (1970): xx.
21. Deleuze and Guattari, *A Thousand Plateaus: Capitalism and Schizophrenia*; Latour, *Reassembling the Social: An Introduction to Actor-Network-Theory*; John Law, "Actor Network Theory and Material Semiotics," in *The New Blackwell Companion to Social Theory* (London: Blackwell, 2009).

Bibliography

Allen, John. 2016. *Topologies of Power: Beyond Territory and Networks*. Abingdon/ New York: Routledge.

Bennett, Jane. 2009. *Vibrant Matter: A Political Ecology of Things*. Durham: Duke University Press.

Bitzer, Lloyd F. 1968. "The Rhetorical Situation." *Philosophy & Rhetoric* 1, no. 1: 1–14.

Bourdieu, Pierre. 1977. *Outline of a Theory of Practice*. Trans. Richard Nice. New York: Cambridge University Press.

Brogan, Walter A. 2012. *Heidegger and Aristotle: The Twofoldness of Being*. Albany: State University of New York Press.

Cassirer, Ernst. 1969. "Mythic, Aesthetic and Theoretical Space." *Man and World* 2, no. 1: 3–17.

Davis, Mike. 2006. *City of Quartz: Excavating the Future in Los Angeles (New Edition)*. New York: Verso Books.

Deleuze, Gilles, and Félix Guattari. 1988. *A Thousand Plateaus: Capitalism and Schizophrenia*. New York: Bloomsbury Publishing.

Foucault, Michel. 1970. *The Order of Things*. New York: Vintage.

———. 1977. *Discipline and Punish: The Birth of the Prison*. New York: Vintage.

———. 1978. *The History of Sexuality, Volume 1: An Introduction*. Trans. Robert Hurley. New York: Random House.

Galloway, Alexander R. 2004. *Protocol: How Control Exists after Decentralization*. Cambridge, MA: MIT Press.

Graham, S. Scott. 2015. *The Politics of Pain Medicine: A Rhetorical-Ontological Inquiry*. Chicago: University of Chicago Press.

Hannah K. Lee. 2016, June. "Can Topology Prevent Another Financial Crash? – Issue 37: Currents." *Nautilus*. http://nautil.us/issue/37/currents/can-topology-prevent-another-financial-crash. Accessed 31 October.

Haraway, Donna. 1988. "Situated Knowledges: The Science Question in Feminism and the Privilege of Partial Perspective." *Feminist Studies* 14, no. 3: 575–599.

———. 1997. *Modest– Witness@Second–Millennium.Femaleman– Meets– Oncomouse: Feminism and Technoscience*. London: Routledge.

Hawhee, Debra. 2002. "Kairotic Encounters." In *Perspectives on Rhetorical Invention*, ed. Janet Atwill and Janice Lauer, 16–35. Knoxville: University of Tennessee Press.

Irigaray, L. 1993. *An Ethics of Sexual Difference*. Ithaca: Cornell University Press.

Kitchin, Rob, and Martin Dodge. 2014. *Code/Space: Software and Everyday Life*. Cambridge, MA: The MIT Press.
Lacan, J., and J.A. Miller. 1998. *The Four Fundamental Concepts of Psycho-Analysis*. New York/London: W.W. Norton.
Latour, Bruno. 2004. "Why Has Critique Run Out of Steam? From Matters of Fact to Matters of Concern." *Critical Inquiry* 30, no. 2: 225–248.
———. 2007. *Reassembling the Social: An Introduction to Actor-Network-Theory*. Oxford: Oxford University Press.
———. 2010. "Steps Toward the Writing of a Compositionist Manifesto." *New Literary History* 41: 471–490.
Law, John. 2009. "Actor Network Theory and Material Semiotics." In *The New Blackwell Companion to Social Theory*, 141–158. London: Blackwell.
Lefebvre, Henri. 1991. *The Production of Space*. Vol. 142. Oxford: Blackwell.
Lury, Celia, Luciana Parisi, and Tiziana Terranova. 2012. "Introduction: The Becoming Topological of Culture." *Theory, Culture & Society* 29, nos. 4–5: 3–35.
Malpas, Jeff. 2008. *Heidegger's Topology: Being, Place, World*. Cambridge, MA: MIT Press.
Muckelbauer, John. 2009. *Future of Invention, The: Rhetoric, Postmodernism, and the Problem of Change*. Albany: State University of New York Press.
Perec, Georges. 1997. *Species of Space and Other Pieces*. Trans. J. Sturrock. Harmondsworth: Penguin.
Pöggeler, Otto. 2005. "Heidegger's Restricted Conception of Rhetoric." In *Heidegger and Rhetoric*, ed. Daniel M. Gross and Ansgar Kemmann, 161–176. Albany: State University of New York Press.
Rotman, Brian. 2012. "Topology, Algebra, Diagrams." *Theory, Culture & Society* 29, nos. 4–5: 247–260.
Sloterdijk, P., and W. Hoban. 2011. *Bubbles: Microspherology*. Cambridge, MA/London: Semiotext(e).
Soja, Edward W. 1996. *Thirdspace: Expanding the Geographical Imagination*. Oxford: Blackwell.
Teyssot, G. 2013. *A Topology of Everyday Constellations*. Cambridge, MA: MIT Press.
Ulmer, Gregory L. 1983. "The Art Object of Post-Criticism." In *The Anti-Aesthetic: Essays on Postmodern Culture*, ed. Hal Foster, 83–95. New York: The New Press.
Vatz, Richard E. 1973. "The Myth of the Rhetorical Situation." *Philosophy & Rhetoric* 6, no. 3: 154–161.

CHAPTER 2

Aristotle's *Topoi* and *Idia* as a Map of Discourse

Sara Rubinelli

The concept of *topos* plays a main role in Aristotle's theory of communication. As shown in the *Topics* and in the *Rhetoric*, *topoi* are at the core of Aristotle's theory of argumentation.[1] As explained in *Topics* 1, 1. 100a 18–19 the aim of the book is "to discover a method by which we shall be able to construct arguments [...] about any problem set before us."[2] The book is a collection of almost 300 *topoi* that can be used by speakers to design arguments on virtually any type of proposition about which there is a difference of opinion. In the *Rhetoric*, having first highlighted the importance of rational persuasion—as based on the use of enthymemes (*Rhetoric* 1, 1. 3)[3]—Aristotle introduces the methodology of the *Topics* in *Rhetoric* 1, 2. 21 as the appropriate one for debating rhetorical issues, that is, issues that are linked to society and are addressed through judicial, deliberative, or epideictic speeches.

This chapter aims to show that the persuasive power of *topoi*, which makes their use valuable in the construction of persuasive speeches, rests on the fact that they play on horizons of discourse that audiences are likely

S. Rubinelli (✉)
Department of Health Sciences and Health Policy, University of Lucerne and Swiss Paraplegic Research, Lucerne, Switzerland

Swiss Paraplegic Research, Nottwil, Switzerland

© The Author(s) 2017

L. Walsh, C. Boyle (eds.), *Topologies as Techniques for a Post-Critical Rhetoric*, DOI 10.1007/978-3-319-51268-6_2

to accept. The conceptualization of *topoi* derives from Aristotle's sharp analysis of the types of content and argumentation strategies that are most likely to receive the consensus of interlocutors. First, the chapter shows that both the *Topics* and the *Rhetoric* present two main, different types of *topoi*: *topoi* in the narrow sense as well as *idia*, that is, *topoi* in the sense of topics, subject matters, or propositions (the Greek *protaseis*). Second, it illustrates how the different *topoi* map and represent horizons of discourse that people are likely to accept. Third, it discusses some implications of considering *topoi* in persuasion and argumentation, with a focus on the ethics of the theory of communication promoted by Aristotle.

Topoi and *Idia*

To understand Aristotle's *topoi* within the framework of topology, here intended as the delineation of discourse that determines the legitimate and illegitimate dialogical moves (see Chap. 1 of this book), it is essential to recall the distinction between the two main types of *topoi* in his works. Specifically, Aristotle distinguishes between *topoi* and *idia* in the famous passage of the *Rhetoric* (1, 2. 1358a 10–21), where he writes:

> I am saying that dialectical and rhetorical syllogisms are those about which we state *topoi*, and these are applicable in common to questions of justice and physics and politics and many different species; for example, the *topos* of the more and the less; for to form syllogisms or speak enthymemes from this about justice will be just as possible as about physics or anything else, although these subjects differ in species. But there are the *idia* that come from the premises of each species and genus [of knowledge]; for example, in physics there are premises from which there is neither an enthymeme nor a syllogism applicable to ethics; and in ethics others not useful in physics. It is the same in all cases. (translation by Kennedy 1991)

The relationship between *topoi* and *idia* is a much debated issue in the literature.[4] To summarize the solution to this debate proposed by Rubinelli, *topoi* and *idia* reflect Aristotle's understanding of the form and content of an argument, respectively.[5]

The *topoi* are strategies of argumentation, specifically argument schemes.[6] They instruct speakers on how to support a standpoint by reflecting on the attribution of a predicate to a subject.[7] According to Aristotle's classification of propositions (*Topics* 1, 5), a predicate can be attributed as *accident*, *property*, *genus*, and *definition*. An accident is a characteristic that does not necessarily belong to a subject (e.g., the quality of being "white" attributed to a "table") (*Topics* 1, 5. 102b 4–9). A property does not show the essence of a thing but rather belongs to it alone (e.g., "to be capable of learning grammar" as the property of "man") (*Topics* 1, 5. 102a 19–22). A genus is part of the essence of something and represents its class in common to other things differing in kind (e.g., "animal" is the genus of both "dog" and "cat") (*Topics* 1, 5. 102a 38–39). A definition is a formula that expresses the essence of a thing (e.g., "virtue of the reasoning faculty" is the definition of "wisdom") (*Topics* 6, 6. 145a 30–31). The *Topics* advises to build an argument by considering whether a predicate can or cannot belong to a subject according to one of the predicates.

Thus, to provide two examples, the following *topos* suggests examining whether the definition of an accident, namely, "envious" can be attributed to "good man":

> Again, to see whether the good man is envious, you must ask, who is 'envious' and what is 'envy'? For if 'envy' is pain at the apparent prosperity of an honest man, clearly the good man is not envious; for then it would be a bad man. (*Topics* 2, 2. 109b 30–33)

The following *topos* suggests examining whether a differentia (an attribute that distinguishes one species from another) has been attributed as the genus of a subject, as this attribution would be incorrect:

> Again, you must see whether your opponent has assigned the differentia as the genus, for example, 'immortal' as the genus of 'God'. For 'immortal' is a differentia of 'living creature'; for some living creatures are mortal and some immortal. It is, therefore, obvious that an error has been committed; for the differentia is never the genus of anything. This is clearly true; for no differentia indicates the essence, but rather some quality, such as 'pedestrian' and 'biped'. (*Topics* 4, 2. 122b 13–18)

In the *Topics*, Aristotle presents a complex system of *topoi*. Indeed, the treatise was composed for dialecticians who were expert in logic and argumentation. Also, the different *topoi* were designed to be used in contexts where speakers needed to be trained in logic,[8] or were engaged in high theoretical discussions, without the appeal of passions, as typical of the political sphere.[9] In the *Rhetoric*, the theory of *topoi* is shortened to the list found in *Rhetoric* 2, 23, which presents a selection of some of the most applicable *topoi* of the *Topics*, and in addition, some argument schemes specifically valuable for building argumentation in a rhetorical context.[10] Thus, for instance, the following *topos* advises orators to look at what turns the mind of the audience in favor of or against something:

> Another [*topos*] that is common both to litigants and deliberative speakers is to look at what turns the mind in favor and what turns the mind against something and for what reasons people both act and avoid action. For these are the factors that if present, impel action [...] (*Rhetoric* 2, 23. 1399b 30–1400a 5)

In order to lead citizens to take a certain action, the orator may, for example, stress that this action is useful for the city or that the eventual damage resulting from the action is inferior to the ultimate advantage that will derive from it.

In the sense of argument schemes, *topoi* are abstract entities: they instruct on how to design an argument and suggest the type of content to use. However, their implementation requires the selection of appropriate content. Thus, in the example above regarding the attribution of the predicate "envious" to the subject "good man," the application of the *topos* requires the use of a general definition of envious and envy.

The section of the *Topics* about the four *organa* (or "instruments") specifically instructs readers on the need to have content available to be able to apply the *topoi* in concrete instances of argumentation. The first *organon* (*Topics* 1, 14. 105a 34–105b 37) suggests that speakers should collect content that is widely shared by people (the *endoxa*, which will be specifically discussed below). The second *organon* instructs on how to apply *topoi* by mastering the several senses of terms, for example, following this illustration by Aristotle:

> If the 'right' is the 'expedient' and the 'honourable', we must try to confirm or demolish both of these terms as applied to the subject under discussion, showing that it is honourable and expedient, or that it is neither honourable nor expedient. (*Topics* 2, 3. 110b 10–15)

The third and fourth *organa* suggest that speakers should know how to recognize the differences and similarities between concepts, again highlighting the value of knowing the exact definitions of things.

Given that dialectical debates may concern virtually any topic (*Topics* 1, 4), the *Topics* focuses on the presentation of the argument schemes and summarizes the need for content and how to find it in the first book, preliminary to the application of the methodology. On the contrary, in the *Rhetoric*, Aristotle places a main emphasis on the content of speeches, that is, on the *idia*, which are the propositions to be used for the construction of arguments in specific contexts. Thus, in the passage of the *Rhetoric* presented at the beginning of this paragraph, Aristotle underlines the universal applicability of *topoi*—due to their formal nature—*versus* the propositions that are context dependent.

Since rhetorical speeches mainly concern judicial, deliberative, and epideictic issues (*Rhetoric* 1, 3), the relevant *idia* are those that relate to the aims of these three types of issues: to show that something is lawful or unlawful (or just or unjust) in judicial rhetoric, advantageous or harmful in deliberation, or honorable or shameful in epideictic rhetoric. Thus, for example, in explaining the *idia* useful in deliberative rhetoric, Aristotle sees in "happiness" the ultimate goal of human action. In *Rhetoric* 1. 5, he lists propositions about what happiness is and which factors contribute to it. Should an orator want to show that a certain action is advantageous, he can use his knowledge of happiness and of its determinants to demonstrate that this action will lead to some form of happiness.

The Foundation of *Topoi* and *Idia*

The value of *topoi* and *idia* as strategies of argumentation rests on the fact that they play on ontological factors and beliefs whose validity or acceptability is shared by people. In other words, *topoi* and *idia* map a horizon of discourse containing what is most likely to be acceptable to people, and thus, most likely to be persuasive when used in argumentation.

The *topoi* in the sense of argument schemes work as inferential rules that can be applied to design arguments in the form of hypothetical syllogisms, the so-called *modus tollens* and *modus ponens*, schematized in the following way:

Modus ponens	*Modus tollens*
If P, then Q	If P, then Q
P	Not Q
Hence Q	Hence not P

These *topoi* instruct on how to build arguments that are valid and thus acceptable from a logical point of view.[11] Hence, for example, the following *topos*...

> Moreover you must derive material from the greater and the less degree [...] One it to see whether the greater degree follows the greater degree, for example, if pleasure is good, and greater pleasure is a greater good, and if to commit injustice is an evil whether to commit a greater injustice is also a greater evil. This commonplace is useful for both purposes; for if the increase of the accident follows the increase of the subject, as described above, it is obvious that it is really an accident of the subject, but if it does not follow it, it is not an accident of it. (*Topics* 2, 10. 114b 37–115a 6)

...assures the logical validity of the following types of arguments:

> If A (good) can be predicated as an accident of B (pleasure), then the increase of A (greater good) can be predicated as an accident of the increase of B (greater pleasure). The increase of A can be predicated as an accident of the increase of B. Hence, A can be predicated as an accident of B;

and

> If A can be predicated as an accident of B, then the increase of A can be predicated as an accident of the increase of B. The increase of A does not belong to the increase of B. Hence, A does not belong to B.

From a persuasion point of view, the validity of an argument is an attribute of its acceptability. This is why the argument schemes in the *Topics* are presented as an appropriate argumentation theory. However, the field of logical fallacies—those arguments that undermine the criteria of validity[12]—shows that the fallacy known as *affirming the consequent* (i.e., if a consequence is said to be true, the antecedent is said to be true) can go unnoticed and be convincing. A typical instance would be: "If one has a fever, he is thirsty. Paul has a fever as he is thirsty."

Each *topos* is informed by principles that support the if–then clause as the basis of the *modus ponens* or *modus tollens*. Braet speaks in this context of a logical and pragmatic level of argumentation.[13] These principles guarantee the correctness of the inference from a pragmatic point of view, for example, in the main proposition of the hypothetical syllogism. Thus, in the example presented above, the main proposition "If A can be predicated as an accident of B, then the increase of A can be predicated as an accident of the increase of B" is a principle that—according to Aristotle—is proven by induction (*Topics* 1, 10. 115a 6).

As explained in detail elsewhere,[14] the *topoi* are based on different types of principles established by aspects including the logic of the predicates, for example:

> For things of which the species is predicated, the genus also must be predicated (*Topics* 4, 1. 121a 25–27);

the ontological dimension, for example:

> Things of which the generations are good things are themselves also good (*Topics* 2, 9. 114a 17–18); and

the relationship between similar things:

> If something is true of one of the like things, it is also true of the other (*Topics* 2, 9. 114b 29–30).

These are all principles stating aspects of the world and the knowledge of it proven by induction (and observation) and are, as such, rather objective. Hence, their role in communication is their value in supporting the construction of arguments that reflect the reality of things.

Having explained the role and value of *topoi* in the domain of human discourse, the *idia* are also presented and examined by Aristotle within this context. More specifically, in the *Topics* Aristotle suggests that speakers apply *topoi* by using the *endoxa* (the generally accepted opinions) as content:

> Generally accepted opinions, on the other hand, are those which commend themselves to all or to the majority or to the wise—that is, to all of the wise or to the majority or to the most famous and distinguished of them. (*Topics* 1, 1. 100b 21–23)

Aristotle distinguishes the opinions of the wise as the privileged horizon of discourse, thus supporting the value of wisdom (and *phronesis* as practical wisdom) in guiding the understanding of things.[15] Moreover, he considers the opinions that everybody supports to be appropriate as a sound basis of argumentation, as universal agreement is indicative of the correctness of a proposition. That which is agreed upon by human beings (the common sense discussed by Livi) has an epistemic value.[16] In addition, it has a value in argumentation and persuasion, as the use of content that people are likely to accept strengthens the application of *topoi* and the overall acceptability of the related argumentation. Returning to the example of the "good man" being "envious," namely:

> Again, to see whether the good man is envious, you must ask, who is 'envious' and what is 'envy'? For if 'envy' is pain at the apparent prosperity of an honest man, clearly the good man is not envious; for then it would be a bad man. (*Topics* 1, 2. 109b 30–33)

it is clear that the extent to which the definition of "envy" is shared by the interlocutor will determine the acceptability of the argumentation. If the interlocutor does not accept the definition, the entire argumentation risks rejection.

In the *Topics*, Aristotle further explains the kinds of *endoxa* that are most appropriate for argumentation and gives examples of each of them. In addition to proper *endoxa*, the other propositions are as follows:

- views that are similar to accepted opinions ("If it is a received opinion that there is a single art of grammar, it might seem to be a received opinion that there is also only one art of flute-playing") (*Topics* 1, 10. 104a 18–19);
- propositions that contradict the contrary of accepted opinions ("If it is a received opinion that one ought to do good to one's friends, it will also be a received opinion that one ought not to do them harm") (*Topics* 1, 10. 104a 22–23);
- propositions in which contraries are stated about contraries ("If we ought to do good to our friends, we ought also to do harm to our enemies") (*Topics* 1, 10. 104a 29–30).

Further, Aristotle values the expertise of people as grantors of knowledge in their context of experience. Indeed, he mentions another specific type of content to be used for its strength in argumentation, namely:

- propositions which accord with expertise ("For example, on questions of medicine one would think as the doctor thinks and in matters of geometry as the geometrician thinks") (*Topics* 1, 10. 104a 35–37).

In *Topics* I, 14. 105a 34–105b 18, Aristotle instructs on how to collect all these *endoxa* as a way to empower dialecticians to be prepared for argumentation on different topics. He emphasizes the value of selecting ideas from written discussions and making descriptions of topics along with quotes from reputable individuals. This is exactly what Aristotle himself did in the *Rhetoric*, in the sections where he introduces all the *idia* relevant for the three genres of speeches.

Between chapters four and fourteen of the first book of the *Rhetoric*, he presents *endoxa* to be used:

- in deliberative rhetoric, with a focus on topics including war and peace, happiness, the good, and the advantages;
- in epideictic rhetoric, with a focus on topics including what is worthy for praise or blame; and
- in judicial rhetoric, with a focus on topics including the wrongdoing and pleasure.

As an example of how the *idia* are presented in the *Rhetoric*, Aristotle speaks about pleasure generally and then discusses some of its nuances, citing the authority of Homer:

> Things hoped for [are pleasurable] that, when present, seem to confer great delights or benefits and to benefit without giving pain. Generally, things that give delight when present [are pleasurable], both when we hope for them and (for the most part) when we remember them. Thus, even anger is pleasurable, as Homer also [said in the verse he] composed about anger,
>
> Which is much sweeter than honey dripping from the comb
>
> For no one feels anger at someone who apparently cannot get revenge, and people are not angry—or are less angry—at those much above them in power. (*Rhetoric*, 1, 11. 1370b 8–17)

Conclusion: On the Ethics of Aristotle's Topology

The *Rhetoric*, a theory of civic discourse—as Kennedy translates the title—was designed to empower orators in public speaking.[17] Aristotle wanted to understand why it is that some people are naturally persuasive. In so doing, however, Aristotle informed a theory of rhetoric by more or less explicitly addressing some of Plato's concerns about this discipline. The same instruments of rhetoric can be used to build sound argumentation or to manipulate an audience. Ultimately, however, the *Rhetoric* contains precepts for engaging in ethical communication. Knowledge of rhetoric is indeed necessary:

> in order that it may not escape our notice what the real state of the case is and that we ourselves may be able to refute if another person uses speech unjustly. (*Rhetoric* 1, 1. 1355a 31–33)

For Aristotle, good rhetoric can be used to strengthen speakers' ability to construct sound arguments for the appropriate transmission of good. While, in fact, "the true and the just are by nature stronger than their opposites" (*Rhetoric* 1, 1. 1355a 20–21), the use of rational persuasion is a precondition for the delivery of truth: indeed, truth is, "capable of apprehension by reason."[18]

Within this framework, the topology proposed by Aristotle is grounded in valid forms of arguments, in sound ontological principles, and in quality content established by the agreement of the wise human beings in general, or experts in specific sectors of the arts and sciences. The *Rhetoric* attempts to reinforce in the city those standards of rationality (of the use of reason), reasonableness (of the correctness of reasoning), and content that, as Brunschwig noted, can be smashed "in the open air of the city."[19] Contrary to the opinion of Brunschwig, however, Aristotle introduces a topology of discourse that respects theoretical distinctions: first and foremost the distinction between the form and matter of an argument and the need to control for the quality of argumentation in both aspects.

Aristotle does not instruct speakers on how to deceive audiences, although he highlights what can be misleading in argumentation. His emphasis on *topoi* and *idia* shows that he defends some standards of discourse and that the treatise targets an essentially honest orator:

> Further, one should be able to argue persuasively on either side of a question, just as in the use of syllogisms, not that we may actually do both (for one should not persuade what is debased) but in order that it may not escape our notice what the real state of the case is and that we ourselves may be able to refute if another person uses speech unjustly. (*Rhetoric* 1, 1. 1355a 31–34)

For Aristotle, rhetoric is the faculty of discovering the means to make probable truths persuasive to an audience (*Rhetoric* 1, 2. 1). He is indeed aware that knowledge alone may not be persuasive:

> even if we were to have the most exact knowledge, it would not be very easy for us in speaking to use it to persuade some audiences. Speech based on knowledge is teaching, but teaching is impossible [with some audiences]. (*Rhetoric* 1, 1. 1355a 24–28)

Thus, he presents a topology of *topoi* and *idia* that captures what is most likely to be accepted by people and which can transmit ideas of value for society.

As highlighted by Rowland and Womack, the topology presented by Aristotle in the *Rhetoric* reflects his constructive and valuable approach to democracy:

> Aristotle assumes that the opinions of the people are generally correct and that the people as a whole are wiser than any individual. He also assumes that truth is more powerful than falsehood and that in free debate truth is likely to prevail. Finally, he assumes that it is the role of the rhetor to lead the people to choose the proper action. Although popular opinion often contains a portion of the truth, it is rarely wholly correct. The rhetor responds to this situation, not by manipulating or deceiving the people, as is common in totalitarian societies, but by using reason and strategies that are consistent with reason to lead the people to dialectically discovered truths.[20]

Following Milanese's analysis of classical rhetoric, Aristotle's topology of *topoi* and *idia* has an epistemic value: embedded in a topology of discourse, and within a theory of persuasion, this Aristotelian approach to truth or falsity rediscovers rhetoric as a powerful discipline to elevate the standards of human communication.[21]

Notes

1. William MA Grimaldi, *Studies in the Philosophy of Aristotle's Rhetoric*, Hermes Enizelschriften 25 (Wiesbaden: Franz Steiner Verlag, 1972); Sara Rubinelli, *Ars Topica: The Classical Technique of Constructing Arguments from Aristotle to Cicero* (Dordrecht, NL: Springer, 2009); Frans H Eemeren van *et al.*, *Handbook of Argumentation Theory* (Dordrecht: Springer, 2014).

2. Aristotle, "Aristotle Ii: Posterior Analytics, Topica," in *Loeb Classical Library* (Cambridge, MA: Harvard University Press, 1960).
3. Grimaldi, *Studies in the Philosophy of Aristotle's Rhetoric*; Myles Fredric Burnyeat, "Enthymeme: Aristotle on the Logic of Persuasion," in *Aristotle's "Rhetoric": Philosophical Essays*, ed. Alexander Nehamas and David J. Furley (New Jersey: Princeton University Press, 1994).
4. See especially Edward Meredith Cope, *An Introduction to Aristotle's Rhetoric* (London: Macmillan, 1897, 1970; repr., 1970); Wilhelm A. de Pater, *Les Topiques D'aristote Et La Dialectique Platonicienne* (Fribourg: St. Paul, 1965); Friedrich Solmsen, *Die Entwicklung Der Aristotelischen Logik Und Rhetorik* (Berlin: Weidmann, 1929); also, William MA Grimaldi, "The Aristotelian Topics," *Traditio* 14 (1958); Brad McAdon, "Probabilities, Signs, Necessary Signs, Idia, and Topoi: The Confusing Discussion of Materials for Enthymemes in the Rhetoric," *Philosophy and Rhetoric* 36, no. 3 (2003).
5. Rubinelli, *Ars Topica: The Classical Technique of Constructing Arguments from Aristotle to Cicero*, 59–70.
6. Antoine Braet, "The Common Topics in Aristotle's Rhetoric: Precursor of the Argumentation Scheme," *Argumentation* 19, no. 1 (2005); Rubinelli, *Ars Topica: The Classical Technique of Constructing Arguments from Aristotle to Cicero*; "Aristotle's Classification of Topoi," *Revue international de philosophie* 270, no. 4 (2014).
7. Jacques Brunschwig, *Aristote: Topiques, Tome I, Livres I–Iv, Texte Établi Et Traduit*, Collection Des Universités De France (Paris: Les Belles Lettres, 1967), XLVI; Oliver Primavesi, *Die Aristotelische Topik: Ein Interpretationsmodell Und Seine Erprobung Am Beispiel Von Topik B*, vol. 94, Zetemata (München: Beck, 1996), 89; Rubinelli, *Ars Topica: The Classical Technique of Constructing Arguments from Aristotle to Cicero*, 8–12.
8. Paul Slomkowski, *Aristotle's Topics* (Leiden-New York-Köln: Brill, 1997), 9–42.
9. Jacques Brunschwig, "Aristotle's Rhetoric as a 'Counterpart' to Dialectic," in *Essays on Aristotle's Rhetoric*, ed. Amélie Oksenberg Rorty (Berkeley, CA: University of California Press, 1996).
10. Rubinelli, *Ars Topica: The Classical Technique of Constructing Arguments from Aristotle to Cicero*, 72–90.
11. Brunschwig, *Aristote: Topiques, Tome I, Livres I–Iv, Texte Établi Et Traduit*, 44; Slomkowski, *Aristotle's Topics*, 99–106.
12. Charles L. Hamblin, *Fallacies* (London: Methuen, 1970).
13. Braet, "The Common Topics in Aristotle's Rhetoric: Precursor of the Argumentation Scheme."

14. Rubinelli, *Ars Topica: The Classical Technique of Constructing Arguments from Aristotle to Cicero*, 23–29.
15. John DG Evans, *Aristotle's Concept of Dialectics* (Cambridge, UK: Cambridge University Press, 1977), 77–95; Luis V. Renon, "Aristotle's Endoxa and Plausible Argumentation," *Argumentation* 12 (1998); Stefano Tardini, "Endoxa and Communities: Grounding Enthymematic Arguments," *Studies in communication sciences* (2005).
16. Antonio Livi, *Il Senso Comune Tra Razionalismo E Scetticismo* (Milano: Massimo, 1992).
17. George A. Kennedy, *Aristotle. A Theory of Civic Discourse* (Oxford: Oxford University Press, 1991).
18. Ibid., 34.
19. Brunschwig, "Aristotle's Rhetoric as a 'Counterpart' to Dialectic," 51.
20. Robert C. Rowland and Deanna F. Womack, "Aristotle's View of Ethical Theory," *Rhetoric Society Quarterly* 15 (1985): 22.
21. Guido Milanese, *Lucida Carmina. Comunicazione E Scrittura Da Epicuro a Lucrezio* (MIlan: Vita e Pensiero, 1989).

Bibliography

Aristotle. 1960. "Aristotle Ii: Posterior Analytics, Topica." In *Loeb Classical Library*. Cambridge, MA: Harvard University Press.

Braet, Antoine. 2005. "The Common Topics in Aristotle's Rhetoric: Precursor of the Argumentation Scheme." *Argumentation* 19, no. 1: 65–83.

Brunschwig, Jacques. 1967. *Aristote: Topiques, Tome I, Livres I–Iv, Texte Établi Et Traduit*, Collection Des Universités De France. Paris: Les Belles Lettres.

———. 1996. "Aristotle's Rhetoric as a 'Counterpart' to Dialectic." In *Essays on Aristotle's Rhetoric*, ed. Amélie Oksenberg Rorty, 34–55. Berkeley: University of California Press.

Burnyeat, Myles Fredric. 1994. "Enthymeme: Aristotle on the Logic of Persuasion." In *Aristotle's "Rhetoric": Philosophical Essays*, ed. Alexander Nehamas and David J. Furley, 3–56. Princeton: Princeton University Press.

Cope, Edward Meredith. 1970. *An Introduction to Aristotle's Rhetoric*. London: Macmillan, 1897, 1970.

de Pater, Wilhelm A. 1965. *Les Topiques D'aristote Et La Dialectique Platonicienne*. Fribourg: St. Paul.

Evans, John D.G. 1977. *Aristotle's Concept of Dialectics*. Cambridge, UK: Cambridge University Press.

Grimaldi, William M.A. 1958. "The Aristotelian Topics." *Traditio* 14: 1–16.

———. 1972. *Studies in the Philosophy of Aristotle's Rhetoric*, Hermes Enizelschriften 25. Wiesbaden: Franz Steiner Verlag.

Hamblin, Charles L. 1970. *Fallacies*. London: Methuen.

Kennedy, George A. 1991. *Aristotle. A Theory of Civic Discourse*. Oxford: Oxford University Press.
Livi, Antonio. 1992. *Il Senso Comune Tra Razionalismo E Scetticismo*. Milano: Massimo.
McAdon, Brad. 2003. "Probabilities, Signs, Necessary Signs, Idia, and Topoi: The Confusing Discussion of Materials for Enthymemes in the Rhetoric." *Philosophy and Rhetoric* 36, no. 3: 223–247.
Milanese, Guido. 1989. *Lucida Carmina. Comunicazione E Scrittura Da Epicuro a Lucrezio*. MIlan: Vita e Pensiero.
Primavesi, Oliver. 1996. *Die Aristotelische Topik: Ein Interpretationsmodell Und Seine Erprobung Am Beispiel Von Topik B*. Zetemata, vol. 94. München: Beck.
Renon, Luis V. 1998. "Aristotle's Endoxa and Plausible Argumentation." *Argumentation* 12: 95–113.
Rowland, Robert C., and Deanna F. Womack. 1985. "Aristotle's View of Ethical Theory." *Rhetoric Society Quarterly* 15: 13–31.
Rubinelli, Sara. 2009. *Ars Topica: The Classical Technique of Constructing Arguments from Aristotle to Cicero*. Dordrecht: Springer.
———. 2014. "Aristotle's Classification of Topoi." *Revue International de Philosophie* 270, no. 4: 433–445.
Slomkowski, Paul. 1997. *Aristotle's Topics*. Leiden/New York/Köln: Brill.
Solmsen, Friedrich. 1929. *Die Entwicklung Der Aristotelischen Logik Und Rhetorik*. Berlin: Weidmann.
Tardini, Stefano. 2005. "Endoxa and Communities: Grounding Enthymematic Arguments. Special Issue on Argumentation in Dialogic Interaction." *Studies in Communication Sciences*: 279–294.
van Eemeren, Frans H., Erik C.W. Krabbe, Francisca Snoeck Henkemans, Bart Verheij, and Jean H.M. Wagemans. 2014. *Handbook of Argumentation Theory*. Dordrecht: Springer.

CHAPTER 3

Topoi and *Tekmēria*: Rhetorical Fluidity among Aristotle, Isocrates, and Alcidamas

Adam W. Cody and Rosa A. Eberly

In Aristotle's *Rhetoric*, *tekmērion* is often translated as "necessary sign," that which need only be gestured toward to be proven. The *tekmērion* is claim and evidence collapsed into a single utterance, refutable only by denying its very existence. Aristotle contrasts *tekmēria* with *topoi*, which are contingent and subject to justification. The work of rhetoric includes the public negotiation not only of *topoi* but also of what can be set apart and denominated as *tekmēria*.

Aristotle did not speak the *Rhetoric* into existence from a shapeless void. By excavating particular rhetorical practices from the discursive environment that preceded the *Rhetoric*, we acknowledge the contestability of what is deemed uncontestable. In the corpora of the Attic orators, the term "*tekmērion*" is used not to denote a necessary sign but to dispute whether or not a proposition is disputable. To preface one's own point as being a *tekmērion* or to

A.W. Cody
Communication Arts and Sciences, Pennsylvania State University, University Park, PA, USA

R.A. Eberly (✉)
Communication Arts and Sciences and English, Pennsylvania State University, University Park, PA, USA

© The Author(s) 2017

L. Walsh, C. Boyle (eds.), *Topologies as Techniques for a Post-Critical Rhetoric*, DOI 10.1007/978-3-319-51268-6_3

refute an interlocutor's point as not a *tekmērion* is to imply that *tekmēria* do indeed require rhetorical justification and are not self-evident.

The word *tekmērion* is related to the more literal *tekmar*, a fixed mark or boundary. The root is also found in a pair of verbs: *tekmērioō* and *tekmairomai*. The former, expressed in the active voice, means "to prove positively." The latter, a deponent of that meddling, muddling, magical middle voice, can mean "to fix by mark" or "to form a judgment." Embedding *topoi* and *tekmēria* in the environment of rhetorical practice from which Aristotle's theoretical terms grew allows for a pivot from the active sense of *tekmērioō*, which frames the self-evident as necessary, absolute, and outside of time, to the middle sense of *tekmairomai*, which recognizes that what is defined as beyond argument is itself justified and judged by argument. Informed by rhetorical practices that preceded it, Aristotle's *tekmērion* provides a helpful instrument for surveying topological chronology. Hence our central claim: topological criticism gains dynamism when it attends to time. Existing scholarship on the *tekmērion* does not attend to the term's temporal qualities. The *tekmērion*, though, describes a *topos* purportedly already settled into necessary and impossible categories, lending it a persuasive force outside of—or without—discursively constructed time.

The quasi-dialogue[1] between Isocrates' *Against the Sophists* and Alcidamas' *About the Writers of Written Speeches*, an argument about the relative temporal characteristics of written versus oral discourse, justifies our claim that topological criticism can be dynamic. While Isocrates wrote to be read and re-read, Alcidamas extemporized winged words not meant to last. Juxtaposed, these texts provide novel insights into the workings of time-in-the-texts and the-texts-in-time. Despite Rickert's claim that *topoi* are stable or static[2] and Cassin's contention that *topoi* were created in opposition to rhetoric in time to disallow contradiction,[3] *topoi* are by nature and consequence dynamic. This chapter seeks to challenge any conception of *topoi* as stable or static by, first, surveying Aristotle on *topoi* and *tekmēria* and by, second, focusing on topical dynamics between Isocrates and Alcidamas in what has been described as "the first written debate"[4] in Western history—an exchange about the relative merits of speaking extemporaneously versus writing in and across periods of time.

On the *Tekmērion*

Although the *tekmērion* has been relatively little treated in recent rhetorical scholarship, it is crucial to a nuanced understanding of rhetorical *topoi*. The *tekmērion*'s rich capacity as a theoretical and critical term describing the construction of time within a text remains unexplored. Grimaldi's "*Semeion*, *Tekmērion*, and *Eikos* in Aristotle's *Rhetoric*" remains a touchstone for academic discussion on the *tekmērion*. Because Grimaldi's article is more concerned with distinguishing two types of probability argument, *semeion* and *eikos*, his treatment of the nonprobabilistic *tekmērion* is incidental to his central claim.[5] Nonetheless, Grimaldi gestures toward the vexing problem of the *tekmērion* in Aristotle when he defines it as a necessary and inevitable relationship of sign and signate, but follows with "*tekmērion* syllogisms will give one a knowledge of the necessity of the conclusion, but they will not give a demonstration of *the reasoned fact* of the conclusion, which is to say a knowledge of *why* it is a necessary conclusion."[6] Although Aristotle's *tekmērion* is supposed to produce certain knowledge, it does so without any passage of discursive time.

Tekmērion has also been used as a marker for tracing the movement of rhetorical thinking into and out of the art of rhetoric. Reguero contends that the term was already in use in ancient Attic oratory and tragedy before it was given theoretical definition by professional rhetoricians.[7] Di Piazza and Piazza identify the various uses and meanings of *tekmērion* in Aristotle's *Rhetoric*, in the anonymous *Rhetoric to Alexander*, and in Antiphon's *Tetralogies*.[8] Like Grimaldi, Di Piazza and Piazza's discussion of *tekmērion* is partnered with attention to *semeion*, as the two terms are typically paired in the ancient corpus. Di Piazza and Piazza demonstrate that the relationship between *tekmērion* and *semeion* found in oratory and rhetorical textbooks is matched in the Hippocratic books on the art of medicine.[9] In ancient medicine, *tekmēria* are the observable symptoms that can prove or disprove a hypothesis with their presence or absence, and *semeia* are the cognitive relationships through which a diagnosis is made.[10]

In addition to comparing the interaction of various simultaneous fields (oratorical practice, drama, medicine, and rhetorical theory) in ancient Greece, modern scholars use the term "*tekmērion*" as a marker of similarity and difference among ancient rhetorical thinkers. Noël shows that rhetoric is portrayed in the Isocratean corpus in a way that corresponds to how it is portrayed in the *Rhetoric to Alexander*.[11] She introduces *tekmērion* as a sample term, a synecdoche of the larger correspondence between the two conceptions of rhetoric. Kraus uses *tekmērion*, among other types of proof-making, to outline the similarities between the *Rhetoric to Alexander* and Aristotle's *Rhetoric*.[12] Rather than identifying similarities among contemporaneous authors, Harari, a specialist in classical logic and proof-making, has made use of the *tekmērion* as a site of contrasting intellectual traditions in the ancient commentaries that followed from the Aristotelian corpus.[13]

Aristotle's *Rhetoric* on *Topoi* and *Tekmēria*[14]

Topoi are introduced in Aristotle's *Rhetoric* as that which rhetorical and dialectical syllogisms are about.[15] These *topoi* are further distinguished as being of common (*koine*) and specific (*idioi*) types. *Topoi* are common insofar as they can be used for syllogisms and enthymemes in a variety of fields, while specific *topoi* are unique to one field only.[16] The concept of the *topos* is not introduced for its own sake, but as a metaphor to make the differences among types of syllogisms and enthymemes more clear.[17] Syllogisms and enthymemes are described in terms of cognitive motion, and for motion to exist, there must be *topoi*, or spaces, to move within and among. The *topos* in *Rhetoric* appears with greater clarity when read within the context of Aristotle's spatial metaphor of reasoning as cognitive motion, a material analog to his dynamic metaphysics of entelechy, the potential motion of each being toward its unique purpose or end.

Just before the passage about common and specific *topoi*, Aristotle identifies the sort of work peculiar to rhetoric. He defines rhetoric in terms of its difference from dialectic and from *technai*, the technical crafts. The differences are threefold. The subject of rhetoric's work is that on which we deliberate, that for which we have no technical crafts, and that which is said to a certain class of hearers (*akroatais*).[18] It is worth noting that the term translated here as "hearers" does not denote "ones who senses a sound." Rather, an *akroatēs* is an active auditor, one who attends mentally to an utterance so that it might be understood. The category of *akroatai* is uniquely rhetorical for two reasons: these auditors cannot take a general view of many things nor follow lengthy reasoning.

Aristotle characterizes the rhetorical audience's faculties of reasoning by means of a spatial metaphor. The first characteristic describes a limitation in the breadth of their vision: two objects appearing too far from each other cannot be seen in the same glance. The second characteristic describes a limitation in the depth of cognitive acuity. The passage describes a line of thought approaching the hearer through space from far away or through time from long ago. For a larger or more complex propositional chain, there is a greater metaphorical distance for the reasoning to cover. When there are too many steps between a conclusion and the premises from which it is derived, rhetoric lends the objects of reasoning an appearance of contiguity.

Aristotle then explains two kinds of reasoning that lend themselves well to rhetorical expression. The first is a syllogism formed "from that which has already been logically concluded."[19] Concluded syllogisms can furnish evidentiary propositions to support further syllogisms. Laid out in full, an argument might appear as a single, extensive syllogism with a conclusion tethered all the way back to originary premises. This kind of argument may be sound and valid but would be too complex to be persuasive. Aristotle describes this kind of unabridged syllogism as *mēkos*, a term denoting great distance, height, or duration. Whether rendered in space or time, Aristotle's use of *mēkos* portrays an argument too unwieldy for use on a human scale. It is the work of rhetoric, then, to obscure some or all of the intermediate points, and to present distant points as contiguous in space and sequential in time.

The focus of the passage then shifts to a discussion of the roles of necessary proof and probabilistic proof in rhetorical and dialectical argument. The term "necessary proof" refers to the matters of argument that cannot be otherwise.[20] The passage up to this point has established that rhetoric is unique in its capacity for persuasion without recourse to the necessary relations of dialectical syllogism. Here, though, Aristotle clarifies that rhetorical argument may also be based on necessary proof, but that this kind is rarer, "for many matters about which there are judgments and speculations can be otherwise."[21] Aristotle does not disallow necessary argument from rhetoric in the way that probabilistic argument is disallowed from dialectic, but questions of human action, deliberation, and inquiry—the fields in which the work of rhetoric can be found—are questions of uncertainty and likelihood. Both dialectical syllogism and rhetorical enthymeme can make use of necessity, but only rhetoric is able to deal persuasively in probabilities.

In contrast with necessary arguments, Aristotle refers to probabilistic argument as *eikos*. "For the probable (*eikos*) is that which happens often, and not crudely, as some define it, but about matters that can be otherwise."[22] Arguments from probable premises require a choice of presentation. The connection between necessary propositions is so strong that one cannot accept the evidence without also accepting the conclusion; they are equivalent. Where there is probability, where matters could be otherwise, the argument's legitimacy relies not on the abstract compulsion of necessary proof, but on the judgments of the rhetors and hearers.

Aristotle then pivots on the distinction between the necessary and the probable, presenting the work of rhetoric in terms of the whole and the part. In this section of the passage, Aristotle returns to the spatial metaphor introduced earlier. This section is, in short, a swarm of prepositions indicating position and movement, nearly devoid of predicate verbs.

Aristotle maps the probabilistic argument onto a flow that emanates from the whole and toward the part with the neatly parallel phrase: "*to katholou pros to kata meros*."[23] This mobile, spatial phrase is repeated with some variation throughout the passage and proves to be crucial to Aristotle's account of necessary and probable proofs. The preposition *pros* indicates lateral motion toward or away from its object, but can also take on the associated meanings of proximity or orientation. This word serves as the fulcrum on which the phrase balances, so that there is an implication of some motile force that proceeds from *to katholou* toward *to kata meros*. These two clauses that flank *pros* each contain another preposition of motion, *kata* (although this is obscured by vowel elision in the word *katholou*, a compound of the words *kata holou*).

The kind of movement represented by the preposition *kata* is downward. The sense of downward motion, as of a duck landing on a pond or a nude descending a staircase, is augmented with notions of proximity and orientation, as might be rendered in English as "beneath" or "downstream." Both prepositions' meanings are modified by the case in which their objects are inflected. If the object of *kata* or *pros* is in the accusative case, the motion or orientation suggested by the preposition is directed toward the object. For example, in this passage, the accusative object *meros* gives the phrase *kata meros* the sense "down toward the part." When these prepositions take the genitive case, however, they indicate direction away from their objects. Just so, the compound *katholou* could be translated as "down from the whole." Both clauses describe downward movement, but Aristotle visualizes the whole in a position of height relative to its parts.

Taken all together, the phrase *to katholou pros to kata meros* can be rendered as "that which comes down from the whole is oriented toward that which goes down toward the part." The whole and the part are not in static counterbalance to each other, but are points of origin and termination on a gradated trajectory. The downward preposition *kata* and the mobile preposition *pros* together present the image of a flow emanating from the whole and descending on the parts. Aristotle's description of reasoning in this passage is replete with motion and fluidity, setting him up to define the role of *topoi* in reasoning as the spaces in and through which this cognitive movement takes place in time.

Significantly, at the same time Aristotle distinguishes types of reasoning by their use of necessary and probabilistic proofs, he turns to the *tekmērion* before exploring *topoi* further. While the *topoi* are sites in which cognitive movements take place, *tekmēria* arrest that movement. Aristotle draws attention to the word's supposed etymology as a border or limit.[24] A *tekmērion* is a proof that, if true, cannot be refuted,[25] or at least, is assumed to have this quality by the *rhetor* making this proof.[26] A chain of reasoning that is considered a *tekmērion* cannot be disputed with a counterargument, but can be refuted only by an outright denial of the truth of its propositions. The *tekmērion* is a limit in that it leaves no room for hearers to choose whether to follow the *rhētor*'s reasoning and adopt it as their own.

When Aristotle returns to the subject of *tekmēria* and probabilistic proofs later in *Rhetoric*, he specifies that the sort of judgments made by a *kritēs*, or critical listener, concerns probabilities.[27] Choosing among alternatives is appropriate only if it is possible for the situation to be otherwise. A *tekmērion* does not admit alternative conclusions if the audience is willing to accept the premises. It is only in probabilistic argument that an audience member can consider the quality of the *rhētor*'s reasoning. An audience member must not only determine the truth of the *rhētor*'s assertions but also accept, modify, or deny the cognitive movements the *rhētor* makes within or among *topoi*. This means that different audience members might determine the argument to be valid or invalid based on different criteria. *Tekmēria* require only the clarity of the audience's recognition; acuity in judgment is appropriate to probabilistic reasoning. Aristotle claims that when people speak of "judging by mental excellence," they are referring to the variable qualities of probabilistic reasoning, not the compelling force of *tekmēria*.[28]

Aristotle adds that *tekmēria*, in addition to being necessary, are also constant.[29] Necessity in reasoning and constancy in chronology are complementary qualities. Regardless of the audience members' good or bad judgment, the *tekmērion*'s conclusion is always justified by the existence of its premises. A necessary relation worthy of the name occurs without variation in any situation. Aristotle cautions that a *kritēs* who believes, rightly or wrongly, that the case at hand rests on a *tekmērion* might consider that the argument is not even subject to judgment.[30] To correct for a miscalculated *tekmērion*, Aristotle suggests reminding the audience of the scope of human judgment. Returning the apparently irrefutable, necessary, eternal *tekmērion* to the realm of the probable and deliberative means drawing attention to the particular events that took place.[31] In addition, these events become more persuasive when they are situated in chronological time. While Aristotle mobilizes *topoi* in theory, Isocrates manifests through his rhetorical practice—writing deliberately for the ages—the inevitability of topical movement across time. Further, for Aristotle, *tekmēria* are proofs whose sources are beyond rhetoric and thus outside of time. For Isocrates, *tekmēria* become *topoi* as rhetorical time extends to include preparatory *diatribē*—the continuous rhetorical engagement that precedes composition.

Isocrates' *Against the Sophists*

Isocrates begins *Against the Sophists* by distinguishing between two different types of people who "try their hand at education."[32] He attributes the poor reputation educators have with laypeople to "those who dare with exceeding thoughtlessness to run their mouths." This idiomatic translation could be rendered more literally as wandering without looking around. From the outset, Isocrates condemns a certain class of educators through a figuration of movement. They wander without deliberate intent and do not look around them for planning or perspective. They are without conscious relationship to the spaces they move through or the relative distance between objects in their field of view.

Isocrates faults these audacious wanderers because they have convinced the laypeople that "those who resolve to be carried away by rest are better than those who wear themselves out (*diatribōntōn*) over philosophy."[33] The commendable and the condemnable types of philosophers are distinguished by the way they spend their time. The wanderers are passively carried away by *rhathumein*, "resting" or "relaxing," a phlegmatic verb of laid-back inertia. Their counterparts, though, are defined by *diatribōntōn*.[34] This active participle refers to a wearing out or rubbing away, as of a stone smoothed by a river or a tool's handle smoothed by continuous usage. The object undergoing *diatribōn* changes over a long duration of time according to the forces affecting it. The word's meaning extends to the passage of time in general and to the changes wrought by repetition. The educators that Isocrates favors wear themselves away with the extended study of philosophy, while the wanderers merely "spend their time on debate."[35]

Isocrates also claims that those educators who spend their time in leisurely debate offer to their students an exact knowledge (*epistēmēs*) of what they must do in life to be prosperous (*eudaimones*).[36] Isocrates denies that this kind of knowledge is a possibility for the human mind. "I consider it clear to all," he writes, "that prescience (*progignōskein*) of the future is not in our nature."[37] The educators are developing a bad reputation not only for how they spend their time, but also for the relationship to time they pretend to develop in their students. This certain knowledge is supposed to be equally applicable in any situation, even in the uncertainty of events yet to come. It is unbounded by any particular duration of time or series of points in time.

Moreover, the educators claimed by Isocrates to offer exact knowledge of the future "lack the ability to speak or counsel even about the present."[38] Eschewing *epistēmē*, Isocrates advises instead "reaching for notions (*doxais*)."[39] The verb *chrōmenous* evokes reasoning and judgment happening through figurative distance. Correct or not, the notions must be approximated by reaching. According to Isocrates, one who reaches for *doxa* rather than cleaving to an unchanging *epistēmē* can hope for better mental unity and uprightness.[40]

Isocrates considers a second kind of educator also unworthy of the name: those who undertake political speaking. The political speechmakers are characterized by their concern with developing a *technē*[41] that can turn students into high-quality speakers regardless of experience or nature.[42] As with the leisurely debaters who advertise an *epistēmē* of prosperity, Isocrates takes issue with the atemporal approach of the teachers of political speaking. Isocrates accuses these teachers of attempting to transmit a "certain knowledge of speeches" that can be applied universally.[43]

To Isocrates, these teachers are inappropriately treating skills of language production as no more complex than learning letters.[44] A "prearranged technique" is an inappropriate approach to language education primarily because it does not account for the difference of situations over time. Learning letters is a false analogy to composing speeches because letters are "immobile" and "always aimed at the same ends."[45] Isocrates objects to these teachers' characterization of language production as fixed and constant. Rather, he considers the most skillful speaker as one whose speech is worthy of the events being discussed and who is able to discover subject matter that others have overlooked.[46] While letters, of necessity, are used in the same ways in all occasions, speechmaking can be variably good or bad depending on its novelty and appropriateness. Isocrates identifies this as the "greatest sign (*sēmeion*) of their dissimilarity."[47] In this alliterative passage, he claims that speeches can have nobility if they are fresh and appropriate (*prepontos*) to the opportunities (*kairōn*) of the moment, while it would be nonsense to speak of the noble usage of individual letters.

Isocrates suggests in rebuttal that original and appropriate responses to situations are developed in a student by careful study over time.[48] He concedes that political acumen and clever speaking, as in any occupation, are at least partially products of natural ability (*euphuesin*). Practical experience (*empeirias*) and training (*paideusis*) can, however, improve on available talents.[49] Isocrates emphasizes that directed and sustained mental exercise is the best method for developing fresh and appropriate responses to periodic opportunities. To Isocrates, the immediate demands of the moment are dependent upon careful practice over a long duration of time. Isocrates specifies that his philosophy does not develop the eloquence that helps to address a situational exigence, nor is it a *technē* that prescribes a universal standard of excellence.[50] Rather, the product of Isocratean education (*paideia*) is plausibility (*epieikeian*), the sensibility and sensitivity to discover what might be worth reaching toward.

Alcidamas' *About the Writers of Written Speeches*

Despite his strong preference for speaking over writing, Alcidamas writes in response to Isocrates a speech critical of those who write speeches. The purpose of Alcidamas' *About the Writers of Written Speeches* is to distinguish written composition from spoken improvisation and to subordinate the former to the latter. Alcidamas centers his distinction on the respective relationships of speaking and writing to time. Alcidamas' key term for spoken improvisation is *parautika*, which refers to that which is nearby or immediate to oneself. The benefits of speaking *parautika*, in the moment, chiefly have to do with minute points in time.[51] According to Alcidamas, it allows one "to make swift use of the enthymemes" and "to follow with a sure aim the opportunity (*kairō*) of the events (*pragmatōn*) and the desires (*epithumiais*) of the people (*anthrōpōn*)." It is this quality, the ability to respond quickly to the demands of the situation, which makes speaking in the moment more attractive to Alcidamas. In contrast, he maligns the practice of "writing over a long period of time."[52] In addition to being unresponsive to immediate circumstances, Alcidamas considers this type of composition easy even for the untrained. It allows for the writer to gather together others' arguments and phrasings, get feedback on the speech, and revise at leisure (*scholēn*). Alcidamas considers language that can be reproduced in the moment without reliance on external sources to be more valuable than writing.

Alcidamas illustrates the relative difficulty of speaking extemporaneously and writing in time with three metaphors from athletics.[53] He compares the improvisational speaker to a person used to lifting heavy weights, or running quickly, or throwing a javelin a far distance. This athlete would, if they need to, be able to lift a lighter weight, run slower, or hit a target at a shorter distance. Someone who writes speeches over a long period of time, then, is the light-lifting, slow-running, near-throwing athlete who would fall short if faced with the more difficult task. Alcidamas' athletic metaphor maps different modes of producing language onto different modes of moving through space. The athletic examples all call for different abilities in movement: moving oneself by running either quickly or slowly, moving an external object either powerfully or weakly, or accurately throwing a missile over a greater or shorter distance.

Alcidamas claims that "in the people's lives, speaking is always and in all matters useful."[54] Whether in the Assembly, the law courts, or "private assemblages," improvised speaking is able to address well the opportunities of the moment. Writing, however, falls short because Alcidamas considers it to occur outside of time. He asks his audience to imagine how ridiculous it would seem for the *rhētor* to compose and learn a speech "while the water-clock in the courts was already flowing (*rhēontos*)."[55] By evoking the gradually emptying water-clock used to allot speaking times, Alcidamas portrays written composition as happening outside the normal flow of time. The writer, portrayed in this way, appears to arrest the progression of time to assemble their thoughts and words. He characterizes this kind of skill as useful only to the tyrant, who can take counsel and form judgments at leisure, calling the people together to listen only when fully ready.[56] In addition to making written speeches appear impractical to the speaking situations of Athenian citizens, Alcidamas casts skill at careful composition with an anti-democratic tint.

Alcidamas then offers what he considers to be the greatest proof (*tekmērion de megiston*) of the superiority of improvised speaking: "those who write speeches for the law courts avoid precision and imitate the interpretations (*hērmeneias*) of the off-hand speakers (*autoschediazontōn*)."[57] That written composers use all their careful preparation to conceal the appearance of careful preparation in their speeches seems a powerful condemnation to Alcidamas. His term here, *autoschediazontōn*, is similar to his other key term for improvised speaking, *parautika*, in that both evoke the nearness to the speaker of the speech materials. *Autoschediazos*, though, is more vivid, referring to that which is within immediate reach. The improvisational speaker appears here as a brawler surprised by a foe, willing to make use of whatever is closest at hand. Alcidamas goes on to claim that sounding off-hand is the "extreme of plausibility" for the written composer.[58] The most probable sounding speech is one that most closely approximates the off-hand style and, from this, Alcidamas claims that improvised speaking is superior to the writing that tries to imitate it.

The different proximities to the *rhētor* that Alcidamas ascribes to writing and improvised speaking figure greatly in his appraisal of them. Alcidamas considers innate qualities more valuable than those augmented by external resources. He condemns the composer of written speeches for keeping their skills with language figuratively farther from them than the spoken improviser does. The writer is only able to display wisdom when equipped with a writing table or book, with the implication that this wisdom is inflated or augmented by the external objects.[59] The wisdom of the improvisational speaker, for Alcidamas, would be more valuable, as it is located in the speaker himself or herself. In addition, just like the wisdom drawn from books, the speaker's facility with words only comes when afforded a duration of time to prepare. Without the preparatory *chronos*, the speaker is "more speechless than the ordinary simpletons."[60] To Alcidamas, a writer's ability to craft a speech is dependent on the writer having the luxury of *chronos*.

Alcidamas expresses this concept in a third way, describing a person who "professes a technical craft (*technas*) of speeches (*logōn*), but seems to have in himself nothing but a small ability (*dunamin*) with speaking (*legein*)."[61] Alcidamas makes crucial distinctions here. The first is between the profession of a technical craft (*technē*) and ability (*dunamis*) in the speaker. The capability that appears emphatically *within the speaker* is made preferable to any kind of art or skill that is developed through training, which takes a long time and can be repeated with some variation for other students. The second crucial distinction is one of grammar. The written composer professes a *technē* of *logōn*—"speeches"—but has no *dunamis* with *legein*—"speaking." To complement the division of *technē* from *dunamis*, Alcidamas marks a clear line between learning about speeches as a concrete set of nouns and having an innate power with the infinitive verb *legein*.

Alcidamas then elaborates on the difficulties faced by a written composer when called upon to speak on short notice. The speaker, Alcidamas writes, "moves with slowness of thought to complete the interpretation."[62] Just as he earlier praised the *parautika* speaker for quickness in assembling enthymemes, Alcidamas compares reasoning and language production to movement. He condemns writing based on this temporal-spatial figuration, as he considers it to take more time to cover less distance. Although Alcidamas does not mention *topoi* specifically, his understanding of reason and language is deeply based in movement, of which time and space are necessary components.

Alcidamas attributes the insufficiencies of written composition to the fickleness of memory. He advises his readers: "So, about off-hand speeches, it is necessary only to have the enthymemes in the mind, while revealing the words at the moment."[63] These speeches are not entirely improvised; Alcidamas considers it possible and worthwhile to separate one's arguments from one's style. Enthymemes are easier to memorize, he says, because there are only a few of them, they figure largely in the speeches, and "enthymemes are each revealed once."[64] It is unclear whether Alcidamas means that any given enthymeme is expressed once *per speech* or one time ever. Nonetheless, he firmly situates argument in one moment of time, without duration or repetition. He contrasts the singularity of enthymemes with words and phrases (*onomata de kai rhēmata*), which differ from each other only a little and must be repeated often.[65] Alcidamas contends that words and phrases are hard to memorize because they vary somewhat as they are repeated over a period of time. Unlike the improvised speaker, someone who composes a written speech must memorize not only the enthymemes, but also the exactly expressed phrases.

Alcidamas leverages the differences in memorization strategy between improvised and written speeches into a difference in temporal construction. He describes how the off-hand speaker and written composer react differently to lapses of memory and unexpected rebuttals. If the improviser forgets one or another enthymeme, the improviser can move on to another point and return to the forgotten enthymeme once it comes back to memory.[66] For that matter, if those who speak without a written manuscript are faced with a refutation that they don't expect, they can respond to or adopt the counterargument without their style appearing irregular or confused.[67] Alcidamas considers a prewritten speech to be incapable of this kind of flexibility. Instead, this kind of speech leaves the speaker at a loss if they forget one of their arguments,[68] or try to add material into their speech at the last minute.[69]

Because the prewritten speech requires long preparation, the precise arrangement and phrasing must be followed as it was planned. This firmly situates the prewritten speech in temporal sequence. Each argument or turn of phrase progresses from the one before it and leads into one that follows. Omitting one link in the chain of reasoning or linguistic expression throws into confusion everything that follows. It is exceedingly difficult, according to Alcidamas, to add something new into the prearranged flow of words and thoughts. The *parautika* speech, though, exists not in the realm of sequence but of simultaneity. The off-hand speaker holds a number of arguments at once, and can deploy them in whatever order seems most appropriate in any given moment. The writer of written speeches prepares for a long duration of time and speaks in sequence; the improvised speaker takes the opportunity of a series of individual moments and speaks one simultaneously held argument at each point in time.

Alcidamas also faults prewritten speeches for being less able to "make use of the desires of the listeners."[70] Specifically though, Alcidamas writes that the speaker should adapt to the listeners' desire for a certain amount of speaking time. He cautions that speakers of prewritten speeches either speak for a longer amount of time than their listeners will stand, or they abandon their speech with the audience still wanting to hear more.[71] Alcidamas considers this such an obstacle to the composer of written speeches because "it is difficult, equal to an impossibility, for human foreknowledge (*pronoian*) to reach the future, to forecast precisely, what orientation (*tropon*) the listeners' minds will have regarding the length (*mēkē*) of what is being said."[72] His primary concern with responding to the desires of the listeners in the moment is that the speaker take an appropriate amount of time with his or her speech. For Alcidamas, it is in the temporal dimension of speaking that human prescience is lacking.

This temporal unresponsiveness gives Alcidamas reason to redouble his attack on written speeches. He next disputes that it is even right to call the written compositions "speeches (*logous*)" at all.[73] Rather, he dismisses them as "images (*eidōla*) and figures (*schēmata*) and imitations (*mimemata*) of speeches" and compares them to bronze statues, stone idols, and drawings of animals.[74] For Alcidamas, speaking is the primary use of language, and writing is the lifeless impostor. Although statues, illustrations, and, by extension, written speeches "give delight in the viewing (*theōrias*), they impart nothing useful to human lives."[75] Alcidamas judges these imitations without use because they are without movement (*akinētōs*).[76] Immobile and inflexible, they cannot be adapted to any opportunities (*kairōn*) that arise. Although he concedes that written speeches may be more beautiful, their inability to move and their inflexibility at various points in time make them without utility for meaningful work.

Toward his conclusion, Alcidamas hedges his claims a bit and admits that there are occasions when careful written composition is appropriate. He concedes that there are times when someone visits him to sample his rhetorical prowess, and he shows them a written piece rather than improvising a speech for them. These visitors are those who have never met him before or who come after a long time has passed.[77] It is only those who visit him frequently, on multiple occasions, who are treated to an extemporaneous demonstration, as they likely already have an accurate appraisal of his abilities. As Alcidamas explains, he recourses to showing written work because "signs (*semeia*) of advancement, which are likely (*eikos*) to be produced in thinking (*dianoia*), are seen most clearly in written speeches."[78] Written speeches show more clearly how facility with language changes over time. Spoken improvisations, in addition to their simultaneous construction and sensitivity to periodic opportunities, are ubiquitous. It is thus difficult to relate them to past or future likenesses. In contrast, Alcidamas claims that the advancement of the soul (*psuchēs*) can be seen in written works as if in a mirror.[79]

At this, Alcidamas reaches a kind of synthesis. He concludes by allowing that the best method of developing a speech is a combination of preparation over a duration of time and adaptation to a period in time. He says that *rhētoroi* should give forethought (*pronoia*) to enthymemes and structure, but should leave the precise wording to improvisation.[80] Although he sees merit in both strategies, he restates that off-hand speaking is a more important skill than careful written composition. He casts the choice that faces aspiring *rhētoroi* as between "nobly using opportunities or precisely speaking words."[81] This parallel structure offers a number of juxtapositions: nobility against precision, using against speaking, and opportunities against words. The choice as Alcidamas frames it is really no choice at all, and he is sure to remind his readers in closing "to speak off-hand always and in all matters."[82]

Conclusion

The spaces between *topoi* and *tekmeria* reveal the ticking-tocking mechanism by which topologies shift across time. What Isocrates and Alcidamas mark implicitly as *tekmēria* in each of their discourses reveals how the *topoi* they share shift in time, illuminating the movement of time within their texts and their texts in time. Alcidamas' choice-that-is-no-choice between nobility and precision exemplifies the enthymemic nature of the *tekmērion*: Alcidamas frames the choice in a way that no one could or would choose precision over nobility. Further, Alcidamas' spatial and mobile metaphors for reasoning and language production are evocative of Aristotle's language regarding arguments moving through and among *topoi*. Ultimately, for Alcidamas the exigencies of particular moments eclipse any concerns with longevity or even ephemeral transcendence. Alternatively, Isocrates eschews universals but refuses to give up on motion upward, on discovering what might be worth reaching toward.

Recognizing the always-limited scope of human judgment, the dynamics of *topoi* can better be discerned in the context of *tekmēria*. The *topos* has movement within it, and the *tekmērion* moves in time. *Tekmēria* delimit, in Lentz's terms, "the world around us," and those limits shift between particular moments, between and among discourses, and over time. Aristotle's discourses on *topoi* and *tekmeria* in the *Rhetoric* manifest an internal dialogue between Platonic and Isocratean influences on his thought. Isocrates' *Against the Sophists* and Alcidamas' *About the Writers of Written Speeches* employ together in dynamic chronology the *topoi* and *tekmēria* preserved by Plato as if in amber. Taken together in the three discourses explicated above, *tekmēria* function as fulcrums between shifting topological tectonics of orality and literacy—and extemporaneity and periodicity—in the dynamics of ancient Greek thought.

Notes

1. Van Hook, Lentz, Schiappa, and Gagarin and Woodruff agree that Alcidamas' discourse was composed in response to Isocrates, but about other matters chronological and topological the authors differ.
2. Rickert, *Ambient Rhetoric*, 44.
3. Cassin, *Sophistical Practice: Toward a Consistent Relativism*, 89.
4. Lentz, *Orality and Literacy in Hellenic Greece*, 128.
5. W. M. A. Grimaldi, "*Semeion, Tekmerion, Eikos* in Aristotle's *Rhetoric*," 383.
6. Ibid., 388.
7. M. Carmen Encino Reguero, "La evolución de algunos conceptos retóricos. *Semeion* y *tekmerion* del s. V al IV a.C.," 402.
8. Salvatore Di Piazza and Francesca Piazza, "The Words of Conjecture: Semiotics and Epistemology in Ancient Medicine and Rhetoric," 15.
9. Ibid., 21.
10. Ibid., 22.
11. Marie-Pierre Noël, "Isocrates and the *Rhetoric to Alexander*: Meaning and Uses of *Tekmerion*," 321.
12. Manfred Kraus, "How to Classify Means of Persuasion: The *Rhetoric to Alexander* and Aristotle on *Pisteis*," 271.
13. Orna Harari, "Simplicius on Tekmeriodic Proofs," 366.
14. Aristotle's accounts of *topoi* differ across his works and thus are dynamic in yet another respect.

15. Aristotle, *Rhetoric* 358a, I.ii.21, trans. A. W. Cody, Greek text from *The "Art" of Rhetoric*, ed. and trans. John Henry Freese, 30.
16. Ibid.
17. Ibid., I.ii.20.
18. Ibid., 1357a, I.ii.12.
19. Ibid., I.ii.13.
20. Ibid., I.ii.14.
21. Ibid.
22. Ibid., I.ii.15.
23. Ibid., 1357b, I.ii.15.
24. Ibid., I.ii.17.
25. Ibid., I.ii.18.
26. Ibid., I.ii.17.
27. Ibid., 1402b, II.xxv.10.
28. Ibid.
29. Ibid.
30. Ibid.
31. Ibid., 1403a, II.xxv.11.
32. Isocrates, *Against the Sophists*, trans., A. W. Cody, Greek text from *Isocrates vol. II*, ed. and trans. George Norlin, 162, section 1.
33. Isocrates, *Against the Sophists*, 162, section 1.
34. Isocrates nearly always uses the word *tropos* rather than *ethos* for "character," and *tropos* is the root within *diatribōntōn*. See Eberly and Johnson in Kennerly and Pfister.
35. Ibid.
36. Ibid., 164, sec. 3.
37. Ibid., 162, sec. 2.
38. Ibid., 166, sec. 8.
39. Ibid.
40. Ibid.
41. Ibid., 168, sec. 9.
42. Ibid., sec. 10.
43. Ibid.
44. Ibid., 170, sec. 12.
45. Ibid.
46. Ibid.
47. Ibid., sec. 13.
48. Ibid., 172, sec. 14.
49. Ibid., sec. 15.
50. Ibid.,176, sec. 21.
51. Alcidamas, *About the Writers of Written Speeches*, trans., A. W. Cody, Greek text from *Alcidamas: The Works and Fragments*, ed. and trans. by J. V. Muir, section 3.

52. Alcidamas, *About the Writers of Written Speeches*, section 4.
53. Ibid., 7.
54. Ibid., 9.
55. Ibid., 11.
56. Ibid.
57. Ibid., 13.
58. Ibid.
59. Ibid., 15.
60. Ibid.
61. Ibid.
62. Ibid., 16.
63. Ibid., 18.
64. Ibid., 19.
65. Ibid.
66. Ibid., 20.
67. Ibid., 24.
68. Ibid., 21.
69. Ibid., 25.
70. Ibid., 22.
71. Ibid.
72. Ibid., 23.
73. Ibid., 27.
74. Ibid.
75. Ibid.
76. Ibid., 28.
77. Ibid., 31.
78. Ibid., 32.
79. Ibid.
80. Ibid., 33.
81. Ibid., 34.
82. Ibid.

Bibliography

Alcidamas: The Works and Fragments, ed. and Trans. J. V. Muir. London: Bristol Classical Press (Duckworth), 2001.

Aristotle. 2006. *The "Art" of Rhetoric*, ed. and Trans. John Henry Freese. Cambridge, MA/London: Harvard University Press.

Cassin, Barbara. 2014. "Topos/Kairos: Two Modes of Invention." In *Sophistical Practice: Toward a Consistent Relativism*, 87–101. New York: Fordham University Press.

Di Piazza, Salvatore, and Francesca Piazza. 2016. "The Words of Conjecture: Semiotics and Epistemology in Ancient Medicine and Rhetoric." *Rhetorica* 34, no. 1: 1–26.

Eberly, Rosa A., and Jeremy David Johnson. 2017. "Isocratean *Tropos* and Mediated Multiplicity." In *Ancient Rhetorics + Digital Networks*, ed. Michele Kennerly and Damien Pfister. University of Alabama Press.

Gagarin, Michael, and Paul Woodruff, ed. 1995. *Early Greek Political Thought from Homer to the Sophists. Cambridge Texts in the History of Political Thought.* Cambridge: Cambridge University Press.

Grimaldi, W.M.A. 1980. "Semeion, Tekmerion, Eikos in Aristotle's Rhetoric." *The American Journal of Philology* 4: 383–398.

Harari, Orna. 2012. "Simplicius on Tekmeriodic Proofs." *Studies in History and Philosophy of Science* 43: 366–375.

Isocrates, *Against the Sophists*, Greek text from *Isocrates* vol. II, ed. and Trans. George Norlin. Cambridge, MA/London: Harvard University Press, 2000.

Kraus, Manfred. 2011. "How to Classify Means of Persuasion: The *Rhetoric to Alexander* and Aristotle on *Pisteis*." *Rhetorica* 29: 263–279.

Lentz, Tony. 1989. *Orality and Literacy in Hellenic Greece*. Carbondale: Southern Illinois University Press.

Macagno, Fabrizio, and Douglas Walton. 2015. "Classifying the Patterns of Natural Arguments." *Philosophy and Rhetoric* 48: 26–53.

Noël, Marie-Pierre. 2011. "Isocrates and the *Rhetoric to Alexander*: Meaning and Uses of *Tekmerion*." *Rhetorica* 29: 319–335.

Reguero, M. Carmen Encinas. 2009. "La evolución de algunos conceptos retóricos. *Semeion* y *tekmerion* del s. V al IV a.C." *Rhetorica* 27: 373–403.

Reinhardt, Tobias. 2007. "Techniques of Proof in 4th Century Rhetoric: Ar. *Rhet.* 2.23-24 and Pre-Aristotelian Rhetorical Theory." *Influences on Peripatetic Rhetoric*, ed. David C. Mirhady.

Rickert, Thomas. 2013. *Ambient Rhetoric: Attunements of Rhetorical Being*. Pittsburgh: University of Pittsburgh Press.

Schiappa, Edward. 1990, Winter. "Did Plato Coin *Rhētorikē*?" *The American Journal of Philology* 111, no. 4: 457–470. http://www.jstor.org/stable/295241

Van Hook, LaRue. 1919. "Alcidamas Versus Isocrates; The Spoken Versus the Written Word." *The Classical Weekly* 12: 89–94. www.jstor.org/stable/4387752

CHAPTER 4

The Shape of Labor to Come

Casey Boyle

"The future is made of the same stuff as the present"

~Simone Weil

This Is Not a Proper Introduction

Henry Ford was a smooth operator. More accurately, Ford labored to build smooth processes for his corporate and manufacturing operations. We often consider his contribution to be developing and implementing uniform automobile production processes as they occur in the place of the factory. Indeed, Ford's achievement of maximizing automobile production through the assembly line's carefully *placed* sequencing of tasks did make for a successful, and profitable, uninterrupted production of consumer products. While Ford capitalized on innovative production processes, the assembly line's innovation was not really the *disruption* that we commonly attribute to it. In fact, David Harvey argues "Ford's organizational and technological innovations were, in many respects, a simple extension of well-established trends."[1] Harvey shows how, far from inventing something from nothing, Ford instead augmented and elongated existing practices that minimized

C. Boyle (✉)
Department of Rhetoric and Writing, University of Texas at Austin,
Austin, TX, USA

© The Author(s) 2017

L. Walsh, C. Boyle (eds.), *Topologies as Techniques for a Post-Critical Rhetoric*, DOI 10.1007/978-3-319-51268-6_4

the cost, time, and labor of producing goods for mass-market distribution and sale. This is to say that Ford's most important contribution was probably the understanding of the importance of not rupturing the connection between production and consumption. Harvey encapsulates Ford's achievement for economic production by writing "[t]he purpose of the five-dollar, eight-hour day … coincidentally meant to provide workers with sufficient income and leisure time to consume the mass produced products the corporations were about to turn out in ever vaster quantities."[2] As Harvey shows, Ford realized that his employees needed to be afforded the money and time to consume the products their labor produced. Where capitalism's biggest critic, Karl Marx, exposed how the system alienated a worker from the products of his (sic) labors,[3] Ford saw an opportunity to reunite the product of the worker's labor with that worker, for a reasonable price. This understanding—that the individual worker needed both money and time to purchase and consume the product he (sic) produced—positioned the ideal capitalist system as enacting a continuous operation between production and consumption, reserving just enough room in that transaction for the capitalist business owner to extract profits from the process that he (sic) initiated. Ford later wished to expand the practical implementation of this assembly line logic into his ideal economic sphere wherein labor became less fixed to any given place and could circulate as freely as did capital's finance. In his 1922 biography, for instance, Ford muses about this very possibility for labor and proposes that:

> There surely ought to be flying squadrons of young men who would be available for emergency conditions in harvest field, mine, shop, or railroad. If the fires of a hundred industries threaten to go out for lack of coal, and one million men are menaced by unemployment, it would seem both good business and good humanity for a sufficient number of men to volunteer for the mines and the railroads. There is always something to be done in this world, and only ourselves to do it. The whole world may be idle, and in the factory sense there may be 'nothing to do.' There may be nothing to do in this place or that, but there is always something to do.[4]

While couched in moral platitudes of providing common benefit to humanity, Ford's musings on worker flexibility betrays his commitment to labor (e.g., human beings) as just another resource to deploy *whenever* needed. Despite the temporal undercurrent to that understanding of labor, it is the "factory sense" along that Ford helped inaugurate that provided labor a firm place, a *topos*, upon and against which political movements banded together to make fair and equitable what often became overbearing production demands. In fact, labor movements rely on just those *topoi* born in the Fordist factory setting, including time of labor, working location, and places for organizing unions. At the foundation for these conditions, for both capital and labor, is the firm belief that the individual composes the smallest unit of the capitalist system. Seen from Ford himself who writes that "Every advance begins in a small way and with the individual. The mass is no better than the sum of the individuals."[5]

The ever-present desire to travel *between* individual places so exemplified by Ford's automobile in American life is once again driving our discussions about fair labor practices. Making news almost daily, ridesharing companies like *Uber* and *Lyft* are testing, if not stretching beyond all recognition, the long-established places of labor practice. While *Uber* and *Lyft* serve as emblems for this new economy, a growing number of related companies also seek to leverage the rapid rise and deployment of locative technologies (e.g., mobile devices and location-aware software applications) to push and redefine what it means to work in an emerging economy whose very moniker twists and stretches in efforts to define itself: the gig economy, on-demand economy, platform economy, networked economy, collaborative economy, or the sharing economy. This final term, the *sharing economy*, has become commonplace for companies whose business model is based on developing software to connect customers to an army of "independent contractors" who then provide the skills and material for that particular product or service (i.e., *Airbnb*, *TaskRabbit*, *InstaCart*, etc.). Optimistically read as "community commerce" or pessimistically as "pyramid-scheme economy," the advances in connecting companies with consumers have undeniably stretched many of the established laws and regulations for workers involved in the services. The sharing economy is fueled by the notion that the people providing the good or services, routinely classified as "independent contractors," are merely sharing unused or overabundant resources in their spare time and are not, in fact, employees worthy of protections under labor laws or regulations. More efficiently than Ford was ever able to accomplish, the move toward sharing economies is making stable and continuous employment a thing of the past and creates what I call *nonemployment*, a condition that increasingly lacks the place, *topos*, that is needed for secure working spaces but whose possibilities must be further developed.

This chapter explores how the rhetorical practice of *topoi* might be productively reconsidered as *topological processes*. As traditionally understood, *topoi* are becoming a less-productive foundation for organizing movements, such as labor, in an emerging sharing economy that functions through smooth processes. Instead, *topological processes*, approaches of analysis and invention that reshape relations for producing new capacities, may serve as a more productive organizing concept for labor discussions. I examine this shift by enacting my own smooth operation between uncommon places.

As I argue elsewhere,[6] rhetorical *topoi* produce not only an assembly line of places for developing common material upon which we rely for discussing content, but also offer an assemblage of mediations that enact a matter of concern. How this dynamic plays out in this chapter is through the topological alignment of traditional scholarly textual resources with more recent online news articles, podcasts, biography, and film references. On the way to reshaping *topoi* as *topology*, the chapter's first section responds to Paolo Virno's claim that contemporary economic and cultural conditions lack special places (*topoi*) and must instead make do with ill-fitting commonplaces (*idioi topoi*). The second section covers the emergence of the "sharing economy" as less a disruption of capitalist economic realities than an intensification. Key to this section will be an examination of a three-episode podcast series that explores precarious conditions underlying labor conditions in the sharing economy. For the third section, I turn to a series of media reports, read alongside Brian Massumi's concept of *ontopower*, that trace the emergence of examples for techniques for *rhetorical topologies*. These examples give us a glimpse into how labor workers are striving to create economies, via the establishment of networks and applications, as the means of their own labor conditions. Ultimately, what this chapter seeks to contribute to rhetorical studies is an argument that topological relations might better respond to systems that structurally displace *topos* as a frame of reference, such as sharing economies. In short, the chapter limns the shape of labor to come.

Stretching Our Common Places

Labor politics have always taken place in thin particular disciplines, at particular locations, and in response to particular circumstances. Take, for instance, the well-known portrayal of an American union organizer in the film *Norma Rae*. The movie's titular character, played by actress Sally

Fields, becomes sick from a textile factory's unsafe working conditions and then leads an effort to form a union of her fellow workers that will work toward safer working conditions. The film's iconic moment occurs when, after being rebuffed several times, Norma writes "UNION" on a cardboard sheet and stands on a table above her laboring coworkers, holding up the sign high above her coworkers. The scene demonstrates a number of common tropes in labor movements, including the need for simple messages, common-ground assembly, and firm stances.

While the factory's workers eventually unionize, this scene encapsulates the extent to which any labor movement must rely on the most basic notion of *place* and, further, demonstrates how labor-oriented political movements more generally depend on the rhetorical notion of *topos.* Like this image of a woman worker holding up a simple sign in the middle of a workplace, the oft-used images so closely associated with labor movements often involve striking workers standing in long lines, holding picket signs and, generally, occupying place. This imagery, exemplified by the iconic image from *Norma Rae*, establishes a useful refrain for labor movements to create an identity and empower their various political encounters. Beyond serving as refrain, the image also indexes the extent to which actual place is protected for labor movements, as seen in the right to post information in common areas of a workplace, the right to gather and discuss union formation, and the right to hold votes to form the said labor union.[7] Each and every labor movement that we have known thus far follows these *commonplace* practices that depend on both shared terminology and literal places common to all for gathering and distributing information and discussion.

The particular style of labor *topoi,* seen in the *Norma Rae* and related labor images, enact what Christa Olson defines as an "embodiable topos." For Olson—examining how *topoi* help perform and advance political goals for indigenous identity in Ecuador— these repeatedly circulated visual forms "[gain] their force by indexing and incorporating available assumptions about the bodies they reference."[8] Olson's notion of *topos* is derived from Carolyn Miller's take on *topoi* as offering the "generative potential of the familiar"[9] as well as Ralph Cintrón's characterization of topoi as "storehouses of social energy."[10] Working from both, the *embodiable topos* not only activates common threads from the repeated image and physical places to enact a figure of common reference but also charges that figure with political force. Olson takes care to explain that *topos* becomes embodiable by involving a "slippage" between "'real'" instances of an identity and the "nexus of circumstances and attributes that signal" that identity.[11] As I read it, the rhetorical force of *topoi* that are embodiable resides in blurring supposedly distinct places of real and imaginary, incorporating both as rhetorical practice. What I would like to argue here is that extending Olson's concept through increasingly multiple mediated and digitally networked labor environments demonstrates that a more broadly construed notion of slippage is needed. *Topoi* no longer just slip but they slide from place to place in a process that enacts the kind of smoothness Ford sought in producing, distributing, and selling automobiles. In this process of sliding, the *topoi* we have relied on cannot be seen as "places" in a traditional sense of firm or stable positions upon which we might work. As I will show further below, when what we might call "the background" of those images is undermined (e.g., closure of the factory), the embodied nature of *topos* becomes a hindrance. This is not to say that these *topoi* are entirely without rhetorical force.

While Olson remarks that her notion of "*embodiable topos*" is "several steps removed from Aristotle's concern with reasoning 'from opinions that are generally accepted'"[12] we would be remiss to sever the latter *topoi*'s connections to Aristotelian versions of the concept. For Aristotle, two inflections of *topoi* could be mobilized. The first, the *koina* or what are commonly referred to as commonplaces, can be applied to general manners of discourse. For instance, using the *koina*, a rhetor might not only draw from a populace's common sayings or known puns but also derive procedures (i.e., general and specific, part and whole, cause and effect) for *locating* a place in a matter of concern upon which to stage an argument. In connection with the commonplaces, Aristotle introduced the *idioi topoi*, or the special places of argument. These *topoi* were reserved for particular kinds of discourses and represented only those places in which specific kinds of arguments were made available. Connected to both argument and cultural forms, both forms of Aristotelian *topoi* enabled citizens to grapple with and engage one another in common and special matters concerning the populace.

As Olson steps from Aristotle, we find current cultural theorists stepping back to Aristotle's *topoi*. In particular, Paolo Virno decries the absence of the special places in today's overly corporate cultures. He writes that:

> The transformation with which we must come to terms can be summarized in this way: in today's world, the 'special places' of discourse and of argumentation are perishing and dissolving, while immediate visibility is being gained by the 'common places,' or by generic logical-linguistic forms which establish the pattern for all forms of discourse.[13]

Virno finds reason to lament our loss of special places because "[t]his means that in order to get a sense of orientation ... we cannot rely on those forms of thought, of reasoning, or of discourse which have their niche in one particular context or another"[14] and must instead found our movements upon one-size-fits-all remedies. We must rely then on "[t]he 'common places'" which "alone, are what exist in terms of offering us a standard of orientation, and thus, some sort of refuge from the direction in which the world is going."[15] And where our world is going, Virno continues, is "the movement to the forefront on the part of the intellect as such, the fact that the most general and abstract linguistic structures are becoming instruments for orienting one's own conduct—this situation, in my opinion, is one of the conditions which define the contemporary multitude."[16] This multitude occurs through slippages among the previously distinct places of individuals, places, politics, labor, and intellect. The abstract conditions in which distant and faraway corporate structures can bend and fold local social practices—as seen in sharing economy companies that circumvent laws through categorical slippage—foreground the need for new conceptual frames that stretch and contort those places that were once all too familiar.

Speaking to this inventive formulation and in response to the loss of special places—these special *topoi* are in many senses linguistic forms but also the institutions and everyday social practices we once relied on to order our relations—Virno proposes the "multitude," an entity whose composition is less the gathering of a number of people than a relational capacity of the group to enact possibilities for itself and others through the very ill-fitting commonplaces. McKenzie Wark helpfully describes Virno's concept as follows: "[a] multitude is not a people, meaning it is a plurality which doesn't become a unity" and, further, "but for the multitude, the collective is not the site of the general will, but rather of the general intellect. The question is thus one of creating forms of post-political, non-representative democracy outside of the state."[17]

Despite its importance for Virno's vision of labor's future, the multitude is without direct precedent. Virno writes that "[w]ith regard to the multitude, we are left, instead, with the absolute lack of codification, with the absence of a clear conceptual vocabulary. But this is a wonderful challenge for philosophers and sociologists, above all for doing research in the Field."[18] Virno further elaborates that "[t]here is a dual movement here, from things to words, and from words to things: this requires the post-Ford multitude. And it is, I repeat, an exciting task."[19] We should consider the invention of new *topoi* practices exciting, but the task is not without hardship. The task requires new forms of labor and a reshaping of commonplaces toward an ongoing topological process.

While labor movements have always relied on the particular, concrete situation to move toward unity, we now must contend with the erosion of those particular, special places upon which we once worked. Just as is shown above by *embodiable topos*, the commonplaces we rely on slip and slide from lived, real conditions and from the circulating *topoi* that index and incorporate them. The concerns that Virno raises are the very problems we are now contending with; namely, the *embodiable topos* we have built up over generations to help regulate and organize labor in productive and safe ways has become as condemned and uninhabitable as the abandoned buildings and factories in Ford's initial manufacturing grounds of Detroit. What do we now do with our *topoi*?

Folding Away the Worker

The three rules for any business once were "Location, Location, Location," but the sharing economy has revised those rules to be "Locating … Locating … Locating." Where businesses once sought out particular physical places that could best take advantage of people's traffic patterns to frequent their stores and consume their services, those businesses now invert the importance of location to flexibly respond to where their customer is. As I mentioned above, the various companies that now lead the sharing economy—*Uber*, *Lyft*, *Instacart*, *TaskRabbit*, *Postmates*, and the like.—all make use of the smartphone and its GPS functions to deliver service to a customer where he or she is located in any given moment. This sharing economy seems simple enough until we unfold its surface to show how the workers have lost their place and have literally become Ford's "squadrons of flying workers." Reported in several venues and in innumerable court cases, the workers for these positions are categorically excluded from being as "employees," considered instead "independent contractors." Given the lack of protections and regulations for the worker in sharing economy industries, I refer to this kind of employment as *nonemployment*. The condition of *nonemployment* involves a worker who must compose for himself or herself the infrastructure (equipment, time, financial management, initial costs, maintenance, and risk) associated with providing the good or service for the company with which that worker is associated. This concept echoes the many laments for the so-called neoliberal economy (wherein individuals are responsible for the security and protections the state once provided) but amplifies those laments in that the *nonemployment* position makes the worker responsible for the equipment and infrastructure of the company. This section attempts to trace the placelessness of the worker in the sharing economy both conceptually and practically through a topological process.

In attending the movement from *location* to *locating* afforded through locative media, this chapter elaborates the cultural shifts Gilles Deleuze discusses in his now-commonplace essay "Postscript on Societies of Control." In the essay, Deleuze traces in Michel Foucault's work how disciplinary society functioned along a series of enclosures not unlike the positions in Ford's automobile factory. An individual would proceed from school, to the barracks, to the factory with occasional stops in the clinic or prison for further disciplinary action as needed. In response to the waning of these disciplinary conditions, the very loss of commonplaces as those firm positions of individuality, Deleuze hones in on the problem of employment, writing that "everyone knows that these institutions are finished, whatever the length of their expiration periods. It's only a matter of administering their last rites and of keeping people employed until the installation of the new forces knocking at the door."[20] Where the disciplinary procedures once laid out ordinal series of progressions for individuals, the control society as sketched by Deleuze folds and stretches that line into a surface of innumerable co-ordinate assemblages that are bent and retooled to respond to *kairotic* emergence of new needs and demands. As Deleuze writes, "Just as the corporation replaces the factory, perpetual training tends to replace the school, and continuous control to replace the examination."[21] Here, we find the constant need to modulate our actions without the benefit of firm ground offered by our disciplines reshaping individuals into *dividuals*. In fact, Deleuze comments directly on labor movements stating that in this emerging society of control: "[o]ne of the most important questions will concern the ineptitude of the unions: tied to the whole of their history of struggle against the disciplines or within the spaces of enclosure, will they be able to adapt themselves or will they give way to new forms of resistance against the societies of control?"[22] Each of the sentences I limn from Deleuze's essay touch on the precarity that labor faces. We no longer have to ask ourselves what might the "rough outlines of these coming forms" of the society of control look like because the sharing economy is well on its way in filling in those details.

A concrete example of the changes conceptually sketched by Virno and Deleuze can be seen, or more accurately heard, in a story reported by Benjamin Walker in his podcast, "The Theory of Everything." In the three-episode podcast arc, "Instaserfs I-III,"[23] Walker explores what is it like to work in the sharing economy "as a worker who would clean bathrooms or what it is like to do someone's grocery shopping for *Instacart*, what it's like to clean someone's bathroom for *HomeJoy*, and I want to know what it is like to drive people around in your car for *Uber*."[24] Explicitly modeled off Barbara Ehrenreich's examination of so-called *unskilled workers*,[25] Walker's "Instaserfs" series (somewhat cheekily) contracts with a worker, Andrew, who spends a full month making money only from sharing economy jobs. The three episodes follow Andrew as he slips in and slides out of different "gigs" ranging from the ridesharing *Lyft*, to the delivery service *Postmates*, to the laundry service *Wash.IO*, and several others.

The first episode begins by Andrew's car failing to meet the automobile standards imposed by *Uber*. A key component in *nonemployment*, as I mentioned above, is that the independent contractor incurs operating costs traditionally taken on by a company. In this case, Andrew was unable to afford the minor repair to a bumper, so *Uber* did not take him on as a contractor. Instead, though, he was able to choose from a range of other ridesharing services and settled on *Lyft* in addition to other delivery services like *Instacart*, *Taskrabbit*, *DoorDash*, and so on. What becomes quite clear immediately is that the sheer abundance of these services, via mobile phone apps, makes it easy for the worker to surf across the various services if one is not a good fit.

As Andrew signs up with these services, he is required to attend "onboarding sessions" or what might have been called "employee orientation" in previous employment conditions. In these sessions, the independent contractors are checked in for necessary infrastructure and related items, namely, "do they have a banking/savings account?" and "do they have friends who might want to sign up?" In addition to those requirements, Andrew is instructed at his *Postmates* onboarding session that as a delivery associate, he is not allowed to touch the customers, not even for a handshake. This depressing fact leads Andrew to propose that "[m]aybe the problem isn't that these companies don't see us as employees. Maybe the real problem is that they don't see us as humans."[26] Andrew goes on to comment that maybe he and the workers are not inhuman but that perhaps the customers are becoming inhuman since they are contracting out the most commonplace experience there is—what it is like to wait in line with other people.

While characterizing the workers as inhuman might be read (or heard) as overly dramatic, it is backed up by some of Andrew's experience working *in-between* the company and its customers. Since the customer's primary interactions with the company are performed through the software app, most transactions effectively cut out the worker. So, where Andrew is paid per delivery, any additional income, like a tip or accolade (customer rating), is done in the privacy of that customer's mobile device. The problem with this arrangement, as Andrew points out in the second episode, is the interactions between him and the customers are too brief to build any rapport to generate a tip and, further, any rating is done out of context. While the former is especially important for income, the latter is vital in that the worker must maintain a fairly high customer rating or be deleted from the app. These interactions demonstrate that, despite being the only human associated with the company, the worker is folded away as much as possible.

It is in the second and third episodes of the *Instaserf* series that the workers' plight through *nonemployment* is rendered apparent. First of all, since the services rely on locative media, both the company and its customers are often afforded the ability to view the progress of the worker, especially that worker's exact location. Andrew comments on this anxiety as he attempts to find parking to pick up food from one establishment for delivery to a customer. He mentions that he worries that the customer and company view him as errant as he repeatedly circles a block searching in vain for legal parking. Related to the parking and having to make one's own resources available for the company, Andrew's time is taken up mainly by waiting in lines for items for pickup and delivery. The lines are often very long and Andrew, humorously, runs into other workers from the same or similar sharing economy companies. In one instance, Andrew notices that a person in front of him in line was wearing a *Doordash* hoodie (another sharing economy company) and the two started up a conversation. They quickly conversed about how they are paid and both worried about parking. The other worker asked Andrew to save his spot while he went to check on his illegally parked car. And while he was still in line, Andrew finds *another* worker from *Doordash* whose account of pay structures contradicts the previous conversation that Andrew had had with the other *Doordash* worker. Andrew then tells his listener, "See, these companies change their policies, like, all the time. There was probably an app update while we were waiting in line."[27] If and when the app updates, the worker's terms and conditions are also *updated*.

Andrew's concern about app updates is typical of working with/for/alongside[28] sharing economy companies. In addition to being constantly under surveillance by the company and its customers, the workers' terms of agreements with that company are changed as frequently as the terms of service agreements that any of us encounter in our own use of software. The changing software app that reconfigures working conditions demonstrates Deleuze's proposition that, in societies of control, the contract upon which the individual would sign and agree is being replaced with the code and password through which "[w]e no longer find ourselves dealing with the mass/individual pair. Individuals have become '*dividuals*,' and masses, samples, data, markets, or '*banks*.'"[29] These *dividuals* are the compositions through which a multiplicity of intensities take shape that may include what we think of as an individual but which is irreducible to the enclosures we typically associate with that unit. Given that the contract and its individualizing signature are no longer applicable for a space of *nonemployment* wherein the very conditions of work occur with the update of an app, the very *topos* of the "independent contractor" and even "individual" is itself reshaped into something far less advantageous for a worker. Such a shift in work dynamics demonstrates not the vertically hierarchical structure of Ford's factory/corporate model, nor the horizontally networked model seen in cooperatives but, instead, a *topological* model whose deformations include both the terminological conditions, as performed by software apps in place of contracts, as well as the cultural practices, as seen in how the workers are folded away from both company and customer. Again, we find the traditional *embodiable topos* of the worker boarded up, locked down, and proclaimed uninhabitable.

Instaserfs shows the emergence of new systems of governance whose topological models of power generate intensities we have not experienced before. Introducing the emerging conditions of *cultural topology*, Celia Lury, Luciana Parisi, and Tiziana Terranova proclaim that "topology is now emerging in the practices of ordering, modeling, networking and mapping that co-constitute culture, technology and science."[30] They argue that culture's *becoming topological* "is not simply that topological thinking develops in the arts, humanities, social and natural science alongside mathematics .... Rather, it is that contemporary culture is itself coming to display a proliferation of surfaces that behave topologically," which can be seen in the "20th-century developments in the gridding of time and space, the proliferation of registers, filing and listing systems, the making and remaking of categories, the identification of populations, and the invention of logistics."[31] While *Instaserfs* shows at a microscale how working conditions can change and be reshaped with the click of a button, Lury *et al.* explain that the spatio-temporal dynamics in play at a global scale are no less flexible. They further write that, "[i]n this multiplication of relations, change is established as constant, normal and immanent, rather than as exceptional or externally produced; that is, forms of economic, political and cultural life are identified and made legible in terms of their capacities for change."[32] This description makes clear how much the sharing economy, and increasingly capitalism at large, relies on topological structures.[33] As the processes inherent in sharing economies and the condition of *nonemployment* now rely on increasing processes of modulation where it previously relied on the disciplined module, so too, as Deleuze notes above, does modulation entrain the worker's activities. From a period of education to constant training, the worker is no longer safely positioned inside an ordinal enclosure but is now *free*, in a pejorative sense, to assume the same flexibility, atemporality, and non-fixity as finance and capital themselves. Thus, *topos* as the commonplace we once held in common is too rigid; rhetorical practice must now find ways to move.

Unfolding Work, Again

A company cannot be clearer in communicating that it considers itself to transcend commonplace business custom and regulation than by naming itself *Uber*. In fact, along with *Uber*, the common action associated with these companies is that they look to rewrite "the rulebook" and to "disrupt." Now, disruption has become a *topos* in the emerging sharing economy for any number of reasons; however, the mode and practice of "disrupting" is far less an actual disruption than its progenitors would like to admit. I turn once more to Henry Ford who, in describing the cultural milieu in which his own life and work took place, states that "[w]e are … in the midst of a change …. A great deal of what we call disturbance—a great deal of the upset in what have seemed to be established institutions—is really but the surface indication of something approaching a regeneration. The public point of view is changing, and we really need only a somewhat different point of view to make the very bad system of the past into a very good system of the future."[34] Again, I remind the reader that this statement was published in 1922; nevertheless, its themes of "disruption" to "established institutions" in order to make "very bad systems of the past" into "very good systems for the future" might as well be statements we attribute to sharing economy companies like *Uber*, *Lyft*, *Airbnb*, and others. These bold and fearless proclamations, made by *bold, proclaiming men* (sic) are, in the end, mere commonplaces. Echoing Ford's own life and work almost a century later, several stories made news over the course of drafting this project that concerned how the sharing economy is "disrupting" the *embodiable topoi* of labor regulation. For instance, *Uber*—and most of these cases and the reporting center on *Uber* as a kind of *topos* that stands in for the entire sharing economy—has been forced to provide the same safety measures for its drivers as regulated cab companies, compelling the ridesharing company to vacate its operations in Austin, Texas[35]; *Uber* has also been to court several times in San Francisco, California to argue against its responsibility to consider its drivers as employees.[36] Perhaps in an inadvertent allusion to Henry Ford's pricing that allowed workers to buy the very products they labored on, *Uber* has also arranged for its potential "independent contractors" to procure low-interest loans for the purchase of acceptable cars that they would then use working with/alongside/for *Uber*.[37] Finally, *Uber* drivers in New York have, in lieu of a labor union that they cannot form, elected to instead form a "solidarity group" to push for labor reform and protections while under contract with *Uber*.[38] In these examples and the additional dozens which I lack the room to cite here, we find that the conditions of *nonemployment* are being rewritten daily but that consistently *Uber* and other sharing economy companies are prevailing in those rewrites.

As the material conditions of labor and production more precisely imitate the flexible dynamics of capital and finance, amplifying prior capitalist relationships. Brian Massumi shows the structure for the neoliberal economy is as much tied to the supposedly irrational forces of affect as it is to rationality. At issue for Massumi is that the individual composes the foundation for this economy. Massumi writes that "[t]he individual subject of interest forming the fundamental unit of capitalist society is internally differentiated, containing its own population of 'minority practices' of contrasting affective tone and tenor .... In other words, there is an *infra-individual* complexity quasi-chaotically agitating within the smallest unit."[39] The possibilities in these tensions are what Massumi terms *ontopower*. What ontopower would do is to redirect intensive forces composing the *dividual* or multitude toward a process of self-priming that, echoing topological invention, leverages prior forms toward inventive new compositions.

Rhetorical practice shares structural similarities with the kinds of intensive redirections and bifurcations we find in Deleuze's *dividual*, Massumi's *ontopower*, Virno's multitude, and mathematical topology.

Expanding on the latter example, Fee-Alexandra Haase proposes that the topological study of change we find in geometry is not only applicable but very much in play in rhetorical commonplaces. Haase posits that, in its basic, linguistic definition, rhetorical figures of speech are any "non-ordinary use of language employed to create an emphasis, amplify a meaning, draw a comparison, and finally to make a saying more persuasive" and that the "[a]ddition, subtraction, transposition, and substitution comprise the four categories of change."[40] Now these techniques match the general sense of typical topological techniques as folding, stretching, twisting, bending, or any other move that does not involve cutting or puncture. While we might initially think that addition or subtraction would violate fundamental topological techniques, I add in the caveat that both addition and subtraction often occur, functionally, through amplification or attenuation as a function of folding. Much like the elasticity in topological encounters, rhetoric's own *rhetoricity* allows for a kind of surface play though rhetorical practice. Haase explains that "[w]hen now using the word 'rhetoricity' as a feature of communication, it comprises the fields of rhetoricity itself, persuasion and the style. On the other hand, rhetoricity is a feature of the channels orality, literacy, visuality, and general mediality."[41] This general mediality—in that mediality, from global positioning satellites to mobile devices, focuses on locative purposes—is very much the process through which the conditions of *nonemployment* occur. That is, as the worker is folded out of the interactions of the sharing economy, demonstrating that a rhetorical topology is at play. Thus, when we call back to the *embodiable topos*, we find that "[r]hetoricity is concerned with the relation between 'words' and 'things' and considers the ubiquitousness of rhetoric as the most valuable feature of rhetoric."[42]

While the prevailing *topos* for categorizing employment has been full-time or part-time employment, each benefiting from varying degrees of protection, the *just-in-time* employment of the sharing economy, or *non-employment*, has reshaped the terms we have used. Right now, at least in the United States, new terms and conditions are being proposed for describing kind of employment made available by companies who fail to claim the workers with/alongside they work. Two related terms, "dependent contractor" or "independent worker," are being considered as possible "third" categories for labor classification. As Arun Sundararajan notes, "many laws will have to slowly be changed to account for the new classification."[43] This shift attempts to recognize the extent to which a semantic problem is also a systemic issue. That is, just as we have *embodiable topoi* that index and incorporate the living conditions as well as the images associated with it, so too does the changing of the terms and categories helps shape the possibilities of working conditions. This is but one technique available in a rhetorical topological shift in that, following Haase, it substitutes a term, *dependent* worker, for another *topos* widely used, *independent* contractor. Of course, further topological co-ordination may also take place.

As Deleuze questions the ability of the unions to invent "new forms of resistance," we might take from that questioning that new forms of resistance might not look at all like the *topos* of resistance—taking stands, going on strike, occupying space—upon which labor has traditionally relied. In fact, such a stand might even look ridiculous in the *codespace* of a sharing economy labor dispute. If there is no break room to post union information, no factory floor to take an iconic stand, no storefront to picket against, then what role do our former *topoi* serve? Following Massumi, Virno, Deleuze, and others, we find that such forms of resistance rely on the individual joining, and propel them into, a collective mass. If the *dividual* and/or multitude are the modulations generating affectual force today, then the very notion of resistance, in a labor movement or otherwise, must be topologically reshaped to account for these new conditions. Henri Bergson argues that the vital processes involved in creative evolution reshape resistance, writing "[w]here it has to direct a movement, it begins by adopting it"[44] and "[w]hen once we have grasped them in their essence by adopting their movement, we understand how the rest of reality is derived from them."[45] Resistance in an era of pervasive citizenship[46] and *nonemployment* might then be a topological technique of *strategic adoption*.

How might *strategic adoption*, as a technique of *rhetorical topologies*, be deployed? From what *topos* might it reshape? One example could be the response of the local tech community in Austin, Texas, when ridesharing companies *Uber* and *Lyft* pulled out of Austin in reaction to regulation disputes against the city. In response to their absence, a group of tech workers, led by Joe Liemandt and Andy Tryba, created *RideAustin*[47] a non-profit ridesharing company that offers workers more stable employment as well as higher pay. *RideAustin* is a rapid response to the vacancy created by *Uber* and *Lyft* that capitalized on the *topos* those companies had initially built. The new entity even explicitly states that its success will be based on the fact that "[t]he previous rideshare companies spent millions of dollars then chose to abandon the 'factory' (drivers & riders) they built. We believe with the backing of the community—*RideAustin* can quickly harness this infrastructure—build the right solution and fill the void and provide a great, safe service."[48] *RideAustin* explicitly adopts and conflates the *topoi* of production and consumption into a *topos* of community. That is, instead of leveraging a shared commonplace to persuade an audience, *RideAustin transposes* the *topoi* topologically, enabling the intensive relationships of the *topoi* to articulate themselves in a different, more productive composition. Further, the effort also reshapes the very relationships between worker and owner, amplifying some of the aspects of that relation while attenuating others.

Most economic aims in a capitalistic system, if not all, are based around an individual's self-interest. The logic and care of that system is handed over to a supposed "invisible hand" that emerges to sort out fair and equitable shares. In a series of rhetorical questions, Massumi primes his reader to understand the danger of our commitment to the individual, self-interested and otherwise, in neoliberal economic conditions. One such question gets at the crux of the matter: "Given the transindividual attunement of my dividual dimension to that of others elsewhere in the economic-relational field of life, why should the decisions moving through me not benefit a distant me, even to my me's disadvantage?"[49] Here, Massumi questions what lies beyond self-interest and pressing on the point that the self, in a control society or as part of and apart from the multitude, is only ever a momentary emergence in a wide topological field of relations. "Nothing," Massumi extends his claim, "divides and multiplies the individual so much as its own relation to the future."[50] This very relation, and all the assemblages that that "individual" is a part of and apart from, exert affective force in a topological manner, stretching and twisting that "individual" in a myriad of ways that, from one angle, appears to be a coherent self. These bifurcations and intensities multiply the individual (read: *dividual*) unit and prepare it for what he calls "ontopower." Massumi traces a series of events—a self-immolation in Tunisia, Egyptian protests, Occupy Wall Street, a student movement in Montreal—to which no localization could be attributed but instead must be "thought-felt." On this global scale, Massumi declares that

> For each new occurrence in the series, there were what Simondon calls 'germinal forms'.... All fields of relation textured by resemblance / similarity / contiguity-based constituencies are riddle with internal structural tensions ... The germinal affection-event catapults the relational field toward a critical point, where it must either fragment into a turf fight between constituencies or pass, as an unspecified whole of flow, into a new constituency .... The effect it produces cannot be predicted: it must be invented.[51]

Key to this passage and the intensive dynamics of any given "individual" (here we should read *topos*) is that which a familiar contour (i.e., resemblance/similarity) is not broken in two but rather stretched to its limit through an agonistic shaping. While the *kairotic* event of topical reshaping may seem to collect into traditional oppositions, the opportunity to invent, to come upon something *new*, can only be enacted by exaggerating continuities found throughout *dividual* compositions.

When examining the instance of *strategic adoption* demonstrated by *RideAustin*, we find that the *topos* of resistance is not the taking of a stand outside/against a movement as much as it is moving that movement toward more productive ends. In this way, it is not the novel categories of "independent worker" or "dependent contractor" being created, but instead the independent contractor being transposed into "dependent capitalist." In contrast with the striking worker who seeks better working conditions through opposition, here a labor movement creates a worker assemblage through the transposed activity of philanthropy. While philanthropy is traditionally a *topos* for elite capitalists whose overflowing coffers dribble in to worthy causes from time to time, this version of philanthropy offered by the *dependent capitalist* is a technique of *rhetorical topology* that reshapes labor resistance. The *dependent capitalist* would be that figure (rarely, if ever, an individual as such) whose existence would be limited to assembling workers and the conditions of working, such as a ridesharing company, without the trappings of the routine and *topos* of capitalism as such. Such a transposition is but one technique for a rhetorical topology of labor, but we might easily find many more shapings by amplifying and attenuating existing relationships toward unforeseen compositions.

A Proper Introduction, Perhaps

How does a labor movement form a lasting union when there is no place, no commonplace, no *topos* upon and through which its workers might assemble? As seen in the sharing economy examples above, the very companies who create the conditions of *nonemployment* are structurally inoculated against many of the traditional *topoi* that have served labor movements for a century. If we are not able to inhabit the *embodiable topoi* we have inherited, then rhetorical practice must twist, stretch, bend, and fold those *topoi* toward inventive relationships through the processural techniques afforded by topology. *Rhetorical topologies* offer practical enactments for conceptual problems and, in reverse, a conceptual engagement for practical concerns. In this Mobius strip-like twist, the practical and conceptual should not be understood as two separate spheres connected only through application but could instead be considered as the very act of labor itself. In what we might refer to as platform capitalism or worker philanthropy, the opportunity to dealienate worker(s) from labor's products becomes less an attempt to reapply labor to product, theory to practice, but to see instead a continuous process that incorporates us all. Our work, our labor involves not taking stands but redirecting movements. *How* we go about doing that task is the very shape of labor to come.

Notes

1. Harvey, David. *The Condition of Postmodernity: An Enquiry into the Origins of Cultural Change*. (Oxford: Blackwell 1990), 125.
2. Ibid., 126.
3. See Marx, Karl, and Ernest Mandel. 1992. *Capital: Volume 1: A Critique of Political Economy*. Translated by Ben Fowkes. Reprint edition. London ; New York, N.Y: Penguin Classics.
4. Ford, Henry, and Samuel Crowther. *My Life and Work*. (Garden City: Doubleday, Page & Co, 1922), 277.
5. Ibid., 277.
6. Casey Boyle. *Rhetoric as A Posthuman Practice*. Manuscript.
7. See "Employee Rights." National Labor Relations Board. Accessed October 15, 2016. https://www.nlrb.gov/rights-we-protect/employee-rights.
8. Olson, Christa J. "Performing Embodiable Topoi: Strategic Indigeneity and the Incorporation of Ecuadorian National Identity." *Quarterly Journal of Speech* 96, no. 3 (2010): 303.
9. Carolyn R. Miller, "The Aristotelian Topos: Hunting for Novelty," in *Rereading Aristotle's Rhetoric*, ed. Alan G. Gross and Arthur E. Walzer (Carbondale: Southern Illinois University Press, 2000): 142.
10. Ralph Cintron, "Democracy and Its Limitations," in *The Public Work of Rhetoric: Citizen-Scholar and Civic Engagement*, ed. John M. Ackerman and David J. Coogan (Columbia: University of South Carolina Press, 2010), 101.
11. Olson, "Performing Embodiable *Topoi*" 304.
12. Ibid., 302.
13. Paolo Virno. *A Grammar of the Multitude: For an Analysis of Contemporary Forms of Life*. (Cambridge, Mass: Semiotext (e), 2003), 36.
14. Ibid., 36.
15. Ibid., 36–37.

16. Ibid., 37.
17. Wark, McKenzie. "As the Veneer of Democracy Starts to Fade…." 2016. *Public Seminar*. (Accessed October 31. http://www.publicseminar.org/2016/07/virno/), n.p.
18. Virno, *A Grammar of the Multitude*, 43–44.
19. Ibid., 44.
20. Deleuze, Gilles. "Postscript on the Societies of Control." *October* 59 (Winter 1992), 4.
21. Ibid., 6.
22. Ibid., 7.
23. See Walker, Benjamin. "Instaserfs (I of III)." *Benjamin Walker's Theory of Everything*. Podcast audio, June 17, 2015. https://toe.prx.org/2015/06/instaserfs-i-of-iii/; "Instaserfs (II of III)." *Benjamin Walker's Theory of Everything*. Podcast audio, June 24, 2015. https://toe.prx.org/2015/06/instaserfs-ii-of-iii/; "Instaserfs (III of III)." *Benjamin Walker's Theory of Everything*. Podcast audio, July 7, 2015. https://toe.prx.org/2015/06/instaserfs-iii-of-iii/.
24. Walker, "Instaserfs (I of III)," 1:08–1:13.
25. See Ehrenreich, Barbara. *Nickel and Dimed: On (Not) Getting By in America*. New York: Picador, 2011.
26. Walker, "Instaserfs (I of III)," 12:38–12:46.
27. Walker, "Instaserfs (II of III)," 8:43–8:49.
28. The problem of defining the relationship between a worker and a company in a sharing economy finds a practical problem of striving to locate just one proposition to match. In what I hope will become more clear throughout the essay is that multiple propositions are needed to best topologically engage the manifold relationships that any given in/dividual has with/in his or her *nonemployment*.
29. Deleuze, "Postscript," 5.
30. Lury, Celia, Luciana Parisi, and Tiziana Terranova. 2012. "Introduction: The Becoming Topological of Culture." *Theory, Culture & Society* 29 (4–5): 3–35, 5.
31. Ibid., 8.
32. Ibid., 4.
33. See Hannah K. Lee June, 2016. "Can Topology Prevent Another Financial Crash? – Issue 37: Currents." *Nautilus*. Accessed October 31. http://nautil.us/issue/37/currents/can-topology-prevent-another-financial-crash.
34. Ford, *My Life and Work*, 267.
35. Taylor, Susan Johnston. 2016. "A World Without Uber." *The Atlantic*, June 16. http://www.theatlantic.com/business/archive/2016/06/a-world-without-uber/487331/.

36. Levine, Dan. 2016. "Uber Drivers Remain Independent Contractors as Lawsuit Settled," *Reuters* April 22. http://www.reuters.com/article/us-uber-tech-drivers-settlement-idUSKCN0XJ07H.
37. See Eric Newcomer & Olivia Zaleski. 2016. "Inside Uber's Auto-Lease Machine, Where Almost Anyone Can Get a Car." *Bloomberg.com*. Accessed October 31. https://www.bloomberg.com/news/articles/2016-05-31/inside-uber-s-auto-lease-machine-where-almost-anyone-can-get-a-car.
38. See Andrew J. Hawkins "Uber drivers in New York can't unionize, but some are forming a labor group." *The Verge*. Accessed October 31, 2016. http://www.theverge.com/2016/5/2/11565778/uber-drivers-new-york-alles-organize-labor-union-settlement.
39. Brian Massumi. *The Power at the End of the Economy*. (Durham: Duke University Press, 2014), 8.
40. Fee-Alexandra Haase. 2008. "Forms Visualizing Communication: Notes Regarding the Parallels of Argumentation between Rhetoric and Geometry as Basic Communication Tools." (SSRN Scholarly Paper ID 1098410. Rochester, NY: Social Science Research Network. https://papers.ssrn.com/abstract=1098410), 3.
41. Ibid., 5.
42. Ibid., 6.
43. Arun Sundararajan. 2016. *The Sharing Economy: The End of Employment and the Rise of Crowd-Based Capitalism*. (Cambridge, Massachusetts: The MIT Press) 184.
44. Henri Bergson. *Creative Evolution*. Trans. Arthur Mitchell. (Mineola: Dover Publications, 1998), 79.
45. Ibid., 401.
46. See Casey Boyle. 2016. "Pervasive Citizenship through #SenseCommons." *Rhetoric Society Quarterly* 46 (3): 269–83.
47. See Ride | Austin. "Bringing Ridesharing back to Austin." 2016. *Ride Austin*. Accessed October 31. http://www.rideaustin.com/.
48. Ibid., FAQ.
49. Brian Massumi. *The Power at the End of the Economy*, 58.
50. Ibid., 9.
51. Ibid., 86.

Bibliography

Bergson, Henri. 1998. *Creative Evolution*. Trans. Arthur Mitchell. Mineola: Dover Publications.

Boyle, Casey. 2016. "Pervasive Citizenship Through #SenseCommons." *Rhetoric Society Quarterly* 46, no. 3: 269–283. doi:10.1080/02773945.2016.1171695.

Cintron, Ralph. 2010. "Democracy and Its Limitations." In *The Public Work of Rhetoric: Citizen-Scholar and Civic Engagement*, ed. John M. Ackerman and David J. Coogan, 98–116. Columbia: University of South Carolina Press.

Deleuze, Gilles. "Postscript on the Societies of Control." *October* 59 (Winter 1992): 3–7.

Ehrenreich, Barbara. 2011. *Nickel and Dimed: On (Not) Getting By in America*. New York: Picador.

Ford, Henry, and Samuel Crowther. 1922. *My Life and Work*. Garden City: Doubleday, Page & Co.

Haase, Fee-Alexandra. 2008. *Forms Visualizing Communication: Notes Regarding the Parallels of Argumentation between Rhetoric and Geometry as Basic Communication Tools*, SSRN Scholarly Paper ID 1098410. Rochester: Social Science Research Network. https://papers.ssrn.com/abstract=1098410.

Harvey, David. 1990. *The Condition of Postmodernity: An Enquiry into the Origins of Cultural Change*. Oxford: Blackwell.

Hawkins, Andrew J. 2016. "Uber Drivers in New York Can't Unionize, but Some are Forming a Labor Group." *The Verge*. http://www.theverge.com/2016/5/2/11565778/uber-drivers-new-york-alles-organize-labor-union-settlement. Accessed 31 October.

Lee, Hannah K. 2016. "Can Topology Prevent Another Financial Crash?—Issue 37: Currents." *Nautilus*, June. http://nautil.us/issue/37/currents/can-topology-prevent-another-financial-crash. Accessed 31 October.

Levine, Dan. 2016. "Uber Drivers Remain Independent Contractors as Lawsuit Settled." *Reuters*, April 22. http://www.reuters.com/article/us-uber-tech-drivers-settlement-idUSKCN0XJ07H.

Lury, Celia, Luciana Parisi, and Tiziana Terranova. 2012. "Introduction: The Becoming Topological of Culture." *Theory, Culture & Society* 29, nos. 4–5: 3–35.

Massumi, Brian. 2014. *The Power at the End of the Economy*. Durham: Duke University Press.

Miller, Carolyn R. 2000. "The Aristotelian Topos: Hunting for Novelty." In *Rereading Aristotle's Rhetoric*, ed. Alan G. Gross and Arthur E. Walzer, 130–148. Carbondale: Southern Illinois University Press.

Newcomer, Eric, and Olivia Zaleski. 2016. "Inside Uber's Auto-Lease Machine, Where Almost Anyone Can Get a Car." *Bloomberg.com*. https://www.bloomberg.com/news/articles/2016-05-31/inside-uber-s-auto-lease-machine-where-almost-anyone-can-get-a-car. Accessed 31 October.

Olson, Christa J. 2010. "Performing Embodiable *Topoi*: Strategic Indigeneity and the Incorporation of Ecuadorian National Identity." *Quarterly Journal of Speech* 96, no. 3: 300–323.

Ride | Austin. 2016. "Bringing Ridesharing Back to Austin." *Ride Austin*. http://www.rideaustin.com/. Accessed 31 October.

Sundararajan, Arun. 2016. *The Sharing Economy: The End of Employment and the Rise of Crowd-Based Capitalism*. Cambridge, MA: The MIT Press.

Taylor, Susan Johnston. 2016. "A World Without Uber." *The Atlantic*, June 16. http://www.theatlantic.com/business/archive/2016/06/a-world-without-uber/487331/.

Virno, Paolo. 2003. *A Grammar of the Multitude: For an Analysis of Contemporary Forms of Life*. Cambridge, MA: Semiotext(e).

Walker, Benjamin. 2015a. "Instaserfs (I of III)." *Benjamin Walker's Theory of Everything*. Podcast Audio, June 17. https://toe.prx.org/2015/06/instaserfs-i-of-iii/.

———. 2015b. "Instaserfs (II of III)." *Benjamin Walker's Theory of Everything*. Podcast Audio, June 24. https://toe.prx.org/2015/06/instaserfs-ii-of-iii/.

———. 2015c. "Instaserfs (III of III)." *Benjamin Walker's Theory of Everything*. Podcast Audio, July 7. https://toe.prx.org/2015/06/instaserfs-iii-of-iii/.

Wark, McKenzie. 2016. "As the Veneer of Democracy Starts to Fade….." *Public Seminar*. http://www.publicseminar.org/2016/07/virno/. Accessed 31 October.

CHAPTER 5

Inventing Mosquitoes: Tracing the Topology of Vectors for Human Disease

Molly Hartzog

Genome databases have become essential tools for collaboration and communication in twenty-first-century science. These are databases where genetic information is stored and shared among members of various scientific communities. Perhaps the most influential database is GenBank, an open access database with more than 250,000 species as of 2012.[1] Sequence data are generally submitted by individual laboratories as a requirement for publication by many academic journals.

In organizing these databases and generating data and metadata, contributors are debating how to best organize a digital analog to the natural world that serves as a reliable communication tool *and* a useful research tool. These two criteria can at times be incompatible. To be a useful research tool, the organization of data should reflect the evolutionary history of a particular species; our understanding of the evolutionary history of a species is often incomplete and, therefore, not stable. For a database to be a reliable communication tool, the system must be stable across space and time. The developers of these databases, then, are put in a rhetorical quandary: How can a database be structured in a way that is both useful for research

M. Hartzog (✉)
Department of English and Foreign Languages, Frostburg State University, Frostburg, MD, USA

© The Author(s) 2017

L. Walsh, C. Boyle (eds.), *Topologies as Techniques for a Post-Critical Rhetoric*, DOI 10.1007/978-3-319-51268-6_5

and reliable for communication? The answer to this is determined by what the database developers see as the community's most significant knowledge about the species, standards for research on these species, and values regarding collaboration and communication practices. In this way, genome databases control the direction of research in a given community.[2] In effect, genome databases provide standard topologies for genetic research.

If genome databases are controlling the direction of research, then they are doing more than housing "raw data" that exist "out there" and can be easily retrieved as one retrieves a folder from a filing cabinet. By thinking of these databases in terms of topology, we can think of these databases in another way, as tools of rhetorical invention that help *generate* scientific knowledge. By scrutinizing genome databases as sites that *generate* arguments rather than *house* data, we can develop a more comprehensive picture of the rhetorical practices of a specific community of researchers. Given the significant role of these databases in twenty-first-century science, it is imperative that we understand the topologies that are embedded in these databases and how they direct the inventive practices of a specific community.

In this chapter, I use a topical method derived from classical rhetoric and contemporary work in rhetoric of science in order to analyze one example of a community-developed genome database. In short, I am creating a meta-topology in order to understand the topology used in genome databases. In other words, topology is both my method of analysis and subject of analysis. Tracing the *topoi* that are present in this database provides a means of mapping what types of arguments are more privileged by the database in its very structure. Through this case study, I hope to provide a concrete methodology for looking at genome databases as a topology for a specific scientific research community. Before delving into these analyses, I provide a cursory review of literature on topical invention in science.

Rhetorical Invention

Especially when talking about science, it is important to recognize the two senses of "invention" captured by the rhetorical canon. First, there is the sense of "invention" as the development of something entirely new, something that did not exist before. Second, there is the sense of "invention" that is synonymous with "discovery."[3] In stages of normal science, scientists seek to make "discoveries" about the natural world. That is, they look to describe phenomena that are already existing in the world but not fully understood. In genetic engineering, these discoveries are then exploited in order to achieve a desired effect (e.g., controlling disease) in the natural world. This is where science crosses the boundary between discovery and invention, in the strictest sense, and scientific knowledge/ discoveries/inventions become patentable.

Generating organism- or genome-specific data and inputting them into a database are activities that harken back to a tradition of natural history epitomized by the work of Carl Linnaeus, who developed a system of classification of the plant and animal kingdoms that is still in use today.[4] Classification and categorization practices are essentially questions of definition: What do we call this organism and others like it? What type of organism is it? This makes classification and categorization into activities of rhetorical invention, as the systematist is essentially *inventing* the species through identifying its boundaries. The practice of definition is one part of stasis theory, which follows four categorical sets of "stases" or "sticking points" in the cooperative investigation of a matter: (1) fact, (2) definition, (3) quality, (4) policy. Questions of fact (e.g., What exists?) are usually addressed and agreed upon before questions of definition (What do we call it?), and so on. Generally speaking, scientific literature deals primarily with the first two stases, fact and definition, with science popularization often moving into the third and fourth stases.[5]

Classification and categorization do not necessarily remain solely within the second stasis. Rather, in order to classify an organism, one must first discover and obtain the organism or fossils (stasis of fact), and before one can subscribe to a particular classification system, one must decide the criteria for assessing the value of a classification system (stasis of quality). Exploring the questions surrounding the stasis of definition would provide an insight into what remains unstable, and what *topoi* are used in the attempt to stabilize these questions.

A speaker (in this case a scientist) assumes a specific audience depending on the *topoi* that are used and the stases that are assumed to already be resolved.[6] Addressing a specific audience requires accepting the established stases, taking these stases as assumed, and adopting the acceptable *topoi* that are accepted as persuasive to this particular community. An exploration of the *topoi* used by a specific community, then, provides a way of understanding how this community collectively thinks and generates new knowledge and technologies.

VectorBase

There are many organizations that pull information from larger databases like GenBank and EMBL to create smaller, organism-specific databases, such as Wormbase, Flybase, and Zebrafish Information Network (ZFIN). Another specialized database, VectorBase houses data on invertebrates (many of which are also included in Flybase) that all share one feature: the ability to transmit disease to humans. This database includes sequence data for 51 different species, including several species of mosquito, tick, fly, louse, snail, and others. These insects transmit diseases such as dengue, malaria, Zika, and Lyme. The ability to transmit disease, or "vector capacity," is the primary *topos* of study for users of this database.

The goal of this database is to "provide web-based resources to the scientific community for organisms considered to be causing or transmitting emerging or re-emerging infectious disease."[7] Improvements to the database have been documented in reports published in database issues of *Nucleic Acids Research* (*NAR*) in 2007, 2009, 2012, and 2015.[8,9,10,11] These reports are published as part of the "database issue" in *NAR*, which has been published annually as the first issue of the year for the past 22 years. These issues include "brief descriptions of bioinformatics databases," and in 2015 included 176 reports, 56 of which described new databases.[12] These issues include reports from a number of different databases spanning all areas of interest of *NAR* readership: chemistry and synthetic biology; computational biology; gene regulation, chromatin and epigenetics; genome integrity, repair and replication; genomics; molecular biology; nucleic acid enzymes; RNA and structural biology.

These reports are different from what one might find usually published in *Nucleic Acids Research.* They do not follow the IMRAD structure typical of a scientific report. Instead, they offer descriptive reports of technical improvements to the interface and web hosting, updates on the types of data and metadata incorporated into the database, funding sources, collaboration, and outreach, and directions for future development. These reports steadily increase in detail and length with each year, beginning with three pages in 2007 and seven pages in 2015. See Table 5.1 for additional details on each report.

The reported improvements to VectorBase have focused on increasing usability for purposes of facilitating community work in genome annotation and comparative analysis. In this way, these reports show directly how the developers of this database envision their targeted audience.

Table 5.1 VectorBase reports

Year published	*Authors*	*Title*	*Word count*	*Page count*
2007	Lawson et al.	VectorBase: a home for invertebrate vectors of human pathogens	1301	3
2009	Lawson et al.	VectorBase: a data resource for invertebrate vector genomics	2225	5
2012	Megy et al.	VectorBase: improvements to a bioinformatics resource for invertebrate vector genomics	2228	6
2015	Giraldo-Calderon et al.	VectorBase: an updated bioinformatics resource for invertebrate vectors and other organisms related with human diseases	3345	7

As I mentioned earlier, identifying one's audience is simultaneously situating oneself within a community. Therefore, to explore how the developers understand their audience (in this case, database users), is, to a certain extent, to explore how they are envisioning themselves as members within that community. Identifying the *topoi* employed by these developers, then, will provide an overview of the beliefs, norms, and values that shape this community and drive interpretation and invention.[13]

To accomplish this survey, I inductively derived a list of expected *topoi* after an initial close reading of these reports, describing what seemed to be recurring special *topoi* throughout the texts, that is, the *topoi* that were specific to this particular rhetorical context. I identified places in each report where I saw these *topoi* operating and defined these *topoi* inductively, through several iterations with a second coder. Portions of the reports omitted from analysis were brief, primarily descriptive, and offered no substantive argument. The vast majority of the text was included for analysis. After identifying where *topoi* occurred and developing a list of expected topoi, their definitions, and examples, I worked with a second coder to further refine the definitions. See Appendix A for definitions and examples of the *topoi* we identified. The texts were segmented by main idea or end of sentence or paragraph. Segments ranged from 1 to 4 sentences and were typically under 100 words. The second coder coded a randomized 30 % of the segmented text that I identified as employing one or more *topoi*. After working with the second coder for three passes over the data, we were able to achieve a negotiated 100 % intercoder reliability. I coded the remainder of the data based on the definitions we developed collaboratively.

The frequencies of the *topoi* I identified in these reports are shown in Table 5.2. This table shows what was most valued by the writers of the report for each year, as well as how those values have shifted with each report. Considering the reports together, "review," "breadth/scope of data," "integration," and "community" were the most commonly evoked *topoi*. We coded as "review" any mention of the data review process in VectorBase, including annotation and re-annotation processes. For example, "Once an annotation is finalized, ***additional analyses are performed*** such as our standard orthology/paralogy relationship predictions (6) and ***cross-referencing to other resources***. This system was ***trialled*** for the R. prolixus and G. morsitans genomes." "Review" increased slightly in the second report (2009), but has decreased in frequency since then.

We coded as "integration" any mention of data being linked, cross-referenced, compared or connected in some way, for example, "***Integration of these data with existing gene sets*** has greatly improved reference gene sets (e.g. An. gambiae) and has led to a new 'patch' build system that uses heuristics to merge manual and automated gene predictions to allow more frequent gene set updates." "Integration" declined in frequency between 2007 and 2012, but then more than doubled in frequency between 2012 and 2015. These results suggest that while the developers may have been initially concerned with reviewing data (in many cases this involved annotation and re-annotation), they have since shifted their focus to the integration of data, which, by my definition, includes "any mention of data being linked, cross-referenced, compared or connected in some way" (see Appendix A for definitions and examples of all coded *topoi*).

Table 5.2 Rates of special *topoi* occurring in VectorBase reports

	2007	*2009*	*2012*	*2015*	*Total*
Efficiency	3% (2)	3% (3)	8% (8)	5% (8)	5% (21)
Collaboration	6% (4)	7% (6)	4% (4)	4% (6)	5% (20)
Review	9% (6)	13% (12)	7% (7)	5% (9)	8% (34)
Consistency	6% (4)	3% (3)	7% (7)	7% (11)	6% (25)
Integration	12% (8)	11% (10)	7% (7)	17% (28)	13% (53)
Search/retrieval	5% (3)	3% (3)	5% (5)	8% (14)	6% (25)
Future work	5% (3)	7% (6)	7% (7)	4% (7)	5% (23)
Community	9% (6)	11% (10)	12% (12)	12% (20)	11% (48)
Past growth	3% (2)	4% (4)	0% (0)	13% (21)	6% (27)
Breadth/scope of data	12% (8)	10% (9)	16% (16)	11% (19)	12% (57)
Total *topoi* identified	66	90	98	168	422

We coded as "breadth/scope of data" any mention or *display* of the breadth, scope, or variety of data included in VectorBase, for example, "VectorBase currently hosts *nine genomes* of which the majority are mosquitoes, reflecting their importance in disease agent transmission. The seven corresponding species are *Anopheles gambiae (three genomes, for the PEST, Mali-NIH, and Pimperena colonies), Aedes aegypti, Culex quinquefasciatus, Glossina morsitans, Ixodes scapularis, Pediculus humanus, and Rhodnius prolixus.*" "Breadth/scope of data" remained relatively consistent across all reports, but slightly higher in 2012 than other years.

Since I developed this list of *topoi* inductively, these would be considered "special *topoi*" in the strict Aristotelian sense. The importance of distinguishing these as "special *topoi*" is simply to indicate that these are the *topoi* that are specific to this particular community of VectorBase developers. What is interesting, then, to note here is not only those *topoi* that are used most frequently but also those that are relatively rare. For instance, while references to "community" are particularly frequent, references to "collaboration" have remained relatively rare. We coded as "community" any mention of users, a community of scientists, work being outsourced to a community, and so on. "Collaboration," which we defined as any mention of two entities collaborating or partnering together to improve VectorBase or benefiting from the collaborative affordances of Vectorbase, occurred much less frequently, but when it did occur, it typically co-occurred with "community."

The contrast between "collaboration" and "community" is surprising, given that one of the major motivations for building databases is the increased ability to collaborate and share data across space and time. Other values one might connect to the use of databases are notably absent. These include references to decreased cost of laboratory procedures, accuracy of data, manual versus automated procedures, and customization. The fact that these *topoi* were only marginally referenced in these reports, while "community" was referenced frequently, suggests a potential point of conflicting values. The developers seem to be concerned primarily with serving a very specific community of users, but are perhaps more concerned with users as consumers rather than as producers of data. Given that "breadth/scope of data" has remained consistently high throughout all reports, and, by my definition, includes "any mention or *display* of the breadth, scope, or variety of data included in VectorBase," it seems that the developers have not indicated that there is a strong need for additional data. Additionally, given that references to "community" have been consistently frequent, the developers must be imagining the contributions of the community in another way than producing more data, since there is no obvious acknowledgment of a *need* for additional data. Then, if we consider that references to "integration" have increased, which involves cross-referencing and linking data for ease of consumption, then the developers seem to be imagining the community as consumers of integrated data rather than producers and contributors of original data. This finding for VectorBase indicates a shift from churning out more and more data to cleaning up and thoroughly reviewing the data that have already been produced. Moving from the wet bench to the computer and working with "big data" requires highly sophisticated information technologies for organizing and mining massive amounts of data. I turn to one such tool in the following section.

Infectious Disease Ontologies

As mentioned earlier, VectorBase's broad target community is researchers of invertebrate vectors of human pathogens. The *Aedes aegypti* mosquito is included for purposes of researching dengue and yellow fever transmission cycles. *Aedes aegypti* data is organized following the standards provided by the Infectious Disease Ontology for Dengue Fever (IDODEN).[14] In addition, as of October 4, 2015, VectorBase houses data on 19 different species of *Anopheles* for the purpose of researching malaria transmission cycles. Like the data on *Aedes aegypti*, these data are organized following an ontology, in this case the Infectious Disease Ontology for Malaria (IDOMAL).

Ribes and Bowker define an ontology as "an information technology for representing specialized knowledge in order to facilitate communication across disciplines, share data or enable collaboration. In a nutshell, they describe the sets of entities that make up the world-in-a-computer, and circumscribe the sets of relationships they can have with each other."[15] While not making this argument explicitly, Ribes and Bowker are implying that ontologies create boundary objects, or entities that facilitate communication and collaboration across boundaries.[16] Observing the development of an ontology in the geosciences, Ribes and Bowker describe the strategies used by participants in this group for developing an ontology. They find that participants were required to communicate across domains of expertise to understand the purpose of an ontology.[17] This process, Ribes and Bowker argue "reconstituted science through the eyes of its data. No longer were data only an individual researchers' raw materials, rather, they became a community resource. But these data were not yet a community resource until they were interoperated, able to move seamlessly across disciplinary, institutional and technical barriers."[18]

This reconstitution of science through data can also be seen in the effort to create ontologies for malaria (IDOMAL) and dengue (IDODEN). For instance, the designers of IDOMAL begin an article reporting on its development by describing the transformation in the approach to disease control:

> The failure of the campaign to eradicate malaria about 40 years ago led, among others, to a widespread notion that this disease can simply not be wiped out. This modified the goals of the majority of malaria workers worldwide towards achieving a mitigation of the problem, rather than seeking a final solution.[19]

Malaria workers rescaled their goal from total eradication of the disease to more reasonable control measures. Part of what prompted this restructuring of goals was the realization that malaria transmission was far more complex than was originally thought, and required input from many different areas of expertise in order to develop a manageable control strategy. This is where information technologies become useful. Ontologies, the IDOMAL designers note, can be used "as an efficient instrument to enhance the impact of IT tools in vector biology and malaria entomology. This efficiency can be achieved by building databases and/or decision support systems driven by wide-ranging ontologies that follow common and established rules."[20] In this case, data become integrated, or "interoperated," to use Ribes and Bowker's term, into a system of support for making decisions on malaria control efforts.[21]

The final step in developing an ontology, according to Ribes and Bowker, is engaging the community for maintenance and use of the database. In this step, participants shift from questions regarding ontology development to questions about the community of users. They argue that in this step, "'the broader community' became important as part of an outreach project: a community which itself had to be engaged and transformed such that they would use and contribute to ontologies ... In order to engage the community in using ontologies and registering their data, members of the community had to understand the value of sharing data, and of ordering them through ontologies."[22]

This emphasis on community that Ribes and Bowker describe is also present in efforts to create an ontology for malaria (IDOMAL) and dengue (IDODEN). The designers of IDOMAL make the case for the use of ontologies explicitly:

> It is apparent that if this kind of data exchange and comprehension by information systems can be achieved, a world-wide malaria eradication campaign would greatly benefit from the adoption of standardized ontologies, which would allow for an extensive data exchange across national boundaries and specific projects.[23]

Furthermore, the designers write that "The aim was to produce a tool that will be useful to the malaria community working towards effectively reducing the global malaria burden."[24] These ontologies re-envision the community that produces this data as a community of users that exchange data. In effect, this ontology re-envisions the "malaria workers" introduced in the opening chapter of the designers' article as the "malaria community" that they discuss in the conclusion. In a sense, through facilitating data integration and exchange, these designers are also designing a community of users.

Similar to the definition provided by Ribes and Bowker, the designers of IDODEN (who include some of the designers of IDOMAL) state that an ontology consists of "definitions of terms in a given domain, as well as, most importantly, the relations that link these terms to each other. Based on the relationships between terms, the parent-children configuration leads to a tree-like format when an ontology is laid out graphically."[25] IDODEN includes 12 relations; IDOMAL includes 11 relations. As shown in Table 5.3, only one relation is included in IDOMAL that is not also included in IDODEN; only two relations are included in IDODEN that are not also included in IDOMAL.

These relations, given their purpose of creating the links between different terms included in IDODEN and IDOMAL, operate quite literally as *topoi*. These relations provide points of departure by linking different concepts through an explicit logical structure. The ontologies that gather and organize these logical structures, IDODEN and IDOMAL, are thus topologies. The user can thus explore the topology of vector capacity by following its relations/*topoi* into familiar and unfamiliar territory. The topology thus provides warrants for generating arguments about the data organized by the ontology. A constantly evolving entity, the topology also continually generates discourse that binds this community of researchers. Exploring these relations through a topological lens provides a way of looking further at the beliefs that drive this community. The relations/*topoi* that the designers chose to include will, to a certain extent, reflect what they believe to be acceptable warrants within this community.

Table 5.3 Relations in IDOMAL and IDODEN with corresponding common topics

Relation	*Example*	*Koinos topos*[a]	*Reasoning family*[a]
Is_a	"Dengue fever" is_a "infectious disease"	12. Parts	Causal
Agent_in	"*Aedes albopictus*" agent_in "dengue transmission"	24. Cause/effect	Causal
Bearer_of	"*Aedes albopictus*" bearer_of "dengue virus"	3. Correlation	Comparative
Happens_during	"Ascites" happens_during "clinical manifestation of dengue"	5. Time	Causal/dimensional
has_role	"*Aedes albopictus*" has_role "infectious agent vector role"	7. Definition and conclusion	Causal
Inheres_in[b]	"Dengue virus seroprevalence" inheres_in "human population"	9. Division	Dimensional
Part_of	"Acquired immunity to dengue" part_of "immunity"	9. Division	Dimensional
Participates_in	"Dengue C protein" participates_in "dengue virion assembly"	3. Correlation	Comparative
Preceded_by	"Vitellogenic stage" preceded_by "previtellogenic development"	5. Time	Causal/dimensional
Precedes[c]	"Pre-oviposition behavior" precedes "egg laying behavior"	5. Time	Causal/dimensional
Realized_by	"Response to visual cue" realized_by "adult vision"	24. Cause/effect	Causal

(*continued*)

Table 5.3 (continued)

Relation	*Example*	*Koinos topos*[a]	*Reasoning family*[a]
Realizes	"Progression of dengue fever" realizes "dengue shock syndrome"	24. Cause/ effect	Causal
Results_in[b]	"Asymptomatic dengue" results_in "convalescence"	24. Cause/ effect	Causal

[a]Numerical identifiers and reasoning families adapted from Walsh (2010)

[b]IDODEN only

[c]IDOMAL only

In order for a warrant to be acceptable, it must adequately reflect a particular belief, value, or norm of a community. The following meta-analysis of these relations tells us what types of arguments are promoted or favored the topologies of VectorBase. By understanding what is favored, I am able to draw conclusions about what is valued in this community.

In order to create a meta-topology of the complex special topics encoded by the IDODEN and IDOMAL ontologies, I turn to the Aristotelian universal topics. These *topoi* have been widely adapted to studies in STEM discourse.[26,27,28] Exploring how these architectonic *topoi* operate in argumentation in specific fields, these scholars were able to achieve a "10,000-foot" view of a community and how it collectively engages in rhetorical invention. Building on Prelli, Walsh added to Aristotle's original catalog nine additional common topics she found present in science, technology, engineering, and mathematics (STEM) research articles.[29,30] These 37 common topics of STEM research were divided into three (overlapping) reasoning families: causal, dimensional, and comparative. In this analysis, I matched each relation in IDODEN and IDOMAL to its Aristotelian counterpart (see Table 5.3). A total of 13 relations are used between the two ontologies. Relations in IDODEN were listed in Mitraka *et al.* and relations from IDOMAL I obtained on October 3, 2015, through a simple command + F search for "relationship" in the browser-based ontology at http://anobase.vectorbase.org/idomal/IDOMAL.obo.[31]

All but four of the relations in IDODEN and IDOMAL reside in the "causal" reasoning family, with some overlap in "dimensional" and "comparative" (see Fig. 5.1).

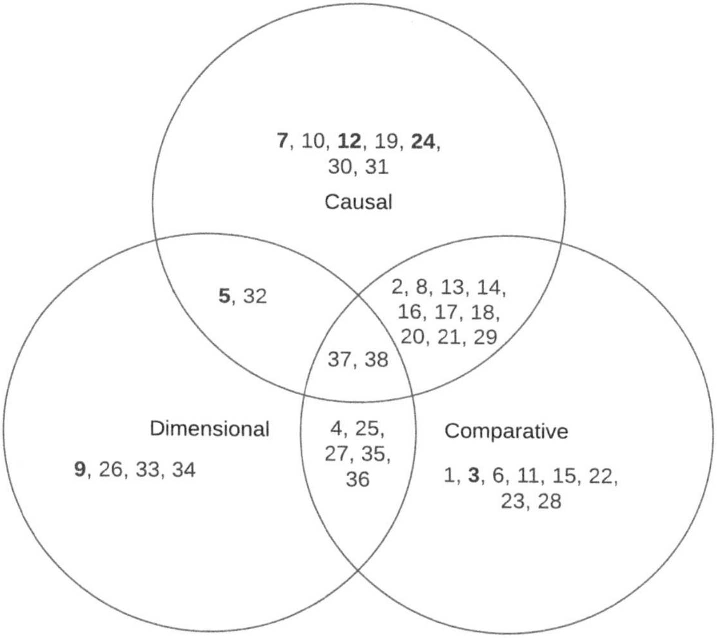

1. Opposites
2. Different forms of the same word
3. **Correlatives**
4. Degree
5. **Time**
6. Turning the tables
7. **Definition and conclusions**
8. Semantics
9. **Division**
10. Induction
11. Precedent
12. **Parts**
13. Evaluation by consequence
14. Catch 22
15. Appearance v. reality
16. Consequence by analogy
17. Similar results must have similar causes
18. Ironic choices
19. Purpose is cause
20. Motives to action or inaction
21. Truth is stranger than fiction
22. Compare/contrast
23. Correcting false impressions
24. **Cause and effect**
25. Search for a better plan
26. Question terms of debate
27. Previous mistakes
28. Meaning of names

Additions from Walsh (2010)

29. Corroboration
30. Methods
31. Assumption
32. Scope
33. Future work
34. Significant anomaly
35. Novelty
36. Unification
37. Consensus

koinoi topoi* in IDOMAL and IDODEN appear in **bold

Fig. 5.1 Common topics in the IDODEN and IDOMAL ontologies by reasoning family

This result is unsurprising given that the primary purpose of these data is to identify (and potentially halt) causes and effects of disease transmission. It is helpful to consider which of these occur most often. Table 5.4 shows the number of occurrences of each unique common topic. In order of most to least occurrences, they are cause/effect, time, division, correlation, parts, and definition and conclusion. What this indicates is that the topology of VectorBase favors greater use of the *topoi* of causality and time, and less use of parts and definition.

This meta-topology of VectorBase licenses some overarching conclusions about the structure of discourse in the malaria-mosquito research community. Walsh argues that the new common STEM *topoi* (#29–37 in Fig. 5.1) indicate a movement in twenty-first century STEM research toward consensus and collaboration, as opposed to difference and conflict.[32] She argues that this relationship between norms and *topoi* should be viewed both ways: that social norms influence the use of certain *topoi* in texts, and texts influence social norms. It is in this way, she argues, that *topoi* "carry knowledge across disciplinary boundaries" by enabling scientists to communicate findings to different stakeholders.[33]

This definition of *topoi* strongly resembles the definition of boundary objects. Boundary objects, as defined by Star and Griesemer, are objects that enable collaboration among different social worlds.[34] These are robust enough to maintain a common identity across boundaries, but flexible enough to adapt to local needs. I do not believe it is the *topoi* that carry knowledge across boundaries, but the boundary objects that carry knowledge, and enable a family of *topoi* to emerge around the object. For example, the "malaria mosquito" functions as a boundary object among different communities of malaria researchers, enabling special *topoi* such as "vector capacity," or the ability to carry and transmit disease, to emerge and serve as a source of rhetorical invention in disease control research.

Table 5.4 Occurrences of unique common topics in IDODEN and IDOMAL

Koinos topos	*Reasoning family*	*Occurrences*
Cause/effect	Causal	4
Time	Causal/dimensional	3
Division	Dimensional	2
Correlation	Comparative	2
Parts	Causal	1
Definition and conclusion	Causal	1

Without the boundary object of the malaria mosquito, the *topos* of "vector capacity" would not be useful, and likely not even exist.

The emphasis on collaboration in VectorBase, IDOMAL, and IDODEN seems to enable and be enabled by the interdisciplinary nature of dengue and malaria research. Given that there are fewer occurrences and thus less emphasis on the *topoi* of correlation, parts, and definition and conclusion in the topology articulated by the ontologies of IDOMAL and IDODEN, these *topoi* seem to be assumed, stable points of agreement. The *topoi* of cause/effect and time are more frequent in these ontologies, suggesting that these are the points of disagreement, or at least highly flexible points, providing a space for invention. If science is being pulled toward collaboration and consensus building, the *topoi* of correlation, parts, and definition and conclusion seem to be the stable points of consensus that *allow for* a more detailed, in-depth exploration in cause/effect and time while still enabling collaboration.

While the use of an ontology-driven database can facilitate collaboration and consensus, there may be some loss in deemphasizing the comparative reasoning family, where novel connections can occur. Other work in rhetorical invention suggests that it is through creative use of metaphor, one form of comparative thinking, that the most innovative arguments can be made.[35] Prelli calls this transpositional thinking. He writes: "By bringing X into relation with Y and viewing X from that vantage point, X displays selective features. X is transformed and is given fresh meaning because Y brings to the fore special details and qualities perhaps previously unforeseen. A transposition thus allows new insights by letting unforeseen relationships come into clear view."[36] By governing the specific relationships a scientist is able to use in this database, the database thus limits the option of creating novel comparisons. While this level of standardization may be useful when considering a database as a communication tool, it is less useful when considering a database as a tool of invention.

Conclusion

This chapter has focused on the development and structure of one genome database, VectorBase. The first analysis looks at a set of published reports written by the developers of VectorBase, where they detail the improvements and updates made to VectorBase from 2007 to 2015. This analysis suggests that developers emphasize the capacity (or potential) of VectorBase to integrate data, and see their intended audience as consumers rather than producers of data. The view of users as consumers rather than producers of data may be a result of the overwhelming amount of data that is undoubtedly housed in VectorBase, shifting the focus from production and collection of additional data to the organization and usability of existing data.

This focus on organization creates an exigence for the Infectious Disease Ontology for Malaria (IDOMAL) and Dengue (IDODEN) that structure part of VectorBase. These ontologies provide a standardized logic for structuring the database into a usable system. In creating these ontologies, the designers are making assumptions about the needs of the research community. These assumptions may be based on past research experience, but they also project future research by highlighting what puzzles are yet to be solved in the current paradigm. These puzzles, however, do not seem to reside in one research jurisdiction; these databases are, essentially, encouraging interdisciplinary research by highlighting interdisciplinary problems. By mapping out knowns and unknowns for researchers, IDODEN and IDOMAL structure the topology of VectorBase, determining the direction of future discourse—and even future research—about the vector capacity of the malaria mosquito.

In Vectorbase, the *topoi* of correlation, parts, and definition and conclusion seem to be the stable points of consensus in this community. This is interesting in itself, given that species are understood to be continually evolving entities, *especially* in the world of entomology; the *topoi* "definition" and "parts," then, one would assume to be highly contentious. However, it seems that these researchers are using the database in a way that stabilizes these species, even if only temporarily, thus allowing a more detailed, in-depth, collaborative exploration in cause/effect and time. The *topoi* of cause/effect and time are more frequent in these ontologies, suggesting that these are the points of disagreement, or at least highly flexible points, providing a space for invention. This makes sense in the context of the ultimate goal of malaria and dengue research: understanding when and how the dengue virus or malaria pathogen is transmitted to humans and manifests as disease.

In this way, genome databases help to build and maintain not just an impressive data set on organisms, but entire research communities. In the case of VectorBase, much of the data is duplicated in other databases, particularly FlyBase. VectorBase, unlike other databases, services an interdisciplinary research community that is focused on a common goal: understanding and controlling vector-borne diseases in humans. Given that this research community includes scientists from traditional disciplines like biology, genetics, epidemiology, virology, and others, their collaborative discourse needs a strong organizing principle if it is to achieve its goal.

The IDOMAL designers, in particular, envision a "malaria community" as a group of interdisciplinary, collaborative workers that focus on "mitigating" the problem of malaria rather than completely eradicating the disease. "Eradication" proved to be a failure. "Mitigation," according to these designers, requires a more collaborative effort than some "eradication" efforts. In providing preestablished lines of reasoning through VectorBase and its organizing ontologies, these designers create a topology that shapes arguments that will be, in theory, widely accepted among this community. However, this topology is dependent on collaboration in order to keep its shape, and collaboration is constrained by the shape of the topology.

In an age where science is becoming more and more interdisciplinary, rhetoricians should adopt methods that look beyond traditional disciplines and instead look at boundary objects and the topologies they generate, to better understand the discourse that develops and sustains these communities. A topological approach would enable a more accurate understanding of the rhetorical practices of a given community, and enable us to better describe interdisciplinary discourses that are focused around a common action, such as controlling disease. This kind of research could potentially contribute back to these interdisciplinary communities by offering them a glimpse of the common beliefs, norms, and values that hold the community together, what might potentially break them apart, and what seems to be most successful in moving toward action. Being an embedded rhetorician in an interdisciplinary program myself, I have found viewing this STEM community through a topological lens helped me to not only navigate the technical vocabulary, but to also identify places for potential intervention to help the technical experts work toward solving global health issues.

Acknowledgments There are far too many who provided stimulating thoughts and direct feedback on this paper to include here, but there are a few who deserve a special acknowledgment. Special thanks to Carolyn R. Miller, for her continuous feedback while this paper was in development, and to Lynda Walsh, for her direction in shaping the paper as part of this collection. I would also like to thank Rene Valdez and Ashley Rose Kelly for their assistance in coding. This project was completed with generous financial and intellectual support from faculty and administrators at the Genetic Engineering and Society Center at North Carolina State University.

Appendix A: VectorBase Report Topoi with Definitions and Examples

1. **Application:** Any mention of a potential marketable or patentable application resulting from VectorBase-related projects

   Example: "The aim of these projects was to better understand the biology of the pathogen through its genome, with the goal of identifying *new therapeutics* and thus shorten the time from therapeutic lead to *marketable product*, a notoriously slow process."
2. **Genome as method:** Any explicit mention of experimental methods, especially the application of VectorBase data to experiment design; additionally, any mention of genome data being used as a vehicle toward learning more about an organism or biological process

   Example: "The availability of the 'Culex' *genome annotation facilitates comparison* of the three main families of mosquitoes (Anopheline, Aedine, and Culicine) with the model dipteran Drosophila melanogaster."
3. **Efficiency:** Any mention of the simplification or ease of workflow within the VectorBase interface, or as a result of using VectorBase, including increased speed or timeliness of work

   Example: "The *simplicity of the submission process* in conjunction with community representative involvement in data quality consistency checks (e.g., does the submitted sequence translate correctly) ensures that any required discussion and error correction *happens in a timely manner.*"
4. **Identifying gap:** Any mention of an area where more knowledge, resources, or data are needed or are lacking

   Example: "A more holistic approach to improving our understanding of these pathogens *needs to include* intermediary vectors where they exist."
5. **Cost:** Any explicit mention of increased or decreased cost of laboratory tools or processes

   Example: "Over the past few years the *cost* of genome sequencing has fallen dramatically making it feasible to sequence the genomes of vectors and complete our knowledge of the triumvirate of species involved in many parasitic diseases."
6. **Funding:** Any explicit mention of funding sources for VectorBase or other entities

Example: "VectorBase is *funded* by the National Institute of Allergy and Infectious disease (NIAID) as part of a group of Bioinformatics Resource Centres (BRCs) (http://www.brc-central.org/) aiming to provide web-based resources to the scientific community for organisms considered to be causing or transmitting emerging or re-emerging infectious disease."

7. **Collaboration:** Any mention of two entities collaborating or partnering together to improve VectorBase or benefiting from the collaborative affordances of Vectorbase; often coincides with "community"

   Example: "VectorBase is involved in all the stages of genome analysis: first-pass annotation of new genome sequences *in collaboration with* the sequencers, re-annotation of existing genome sequences and submission of these data sets to the public nucleotide databanks."

8. **Submission:** Any mention of any portion of the process of submitting data to VectorBase

   Example: "*Data can be submitted* to the VectorBase Population Biology Resource via spreadsheet forms using open source tools to assist with formatting and ontology term selection (ISA-Tab (27) and Phenote, http://www.phenote.org). Genotypes are submitted to the variation resource in standard VCF format (5)."

9. **Review:** Any mention of the data review process in VectorBase, including annotation and re-annotation processes

   Example: "Once an annotation is finalized, *additional analyses are performed* such as our standard orthology/paralogy relationship predictions (6) and *cross-referencing to other resources.* This system was *trialed* for the *R. prolixus* and *G. morsitans* genomes."

10. **Accuracy:** Any mention of the accuracy or precision of data or technical process, or corrections to those data or techniques

    Example: "These data include *corrections* of gene structures and relevant metadata such as gene symbols and citations."

11. **Consistency:** Any mention of efforts toward organizing, storing, ordering data in a consistent manner, including any mention of "ontologies"

    Example: "The collection of experimental and sample-related metadata has been aided through our development of *ontologies and controlled vocabularies* for vector-specific data, such as field-associated samples, pathogen transmission and insecticide resistance"

12. **Integration:** Any mention of data being linked, cross-referenced, compared or connected in some way

    Example: "*Integration of these data with existing gene sets* has greatly improved reference gene sets (e.g., An. gambiae) and has led to a new 'patch' build system that uses heuristics to merge manual and automated gene predictions to allow more frequent gene set updates"
13. **Search/retrieval:** Any explicit mention of the process of data mining or searching or retrieving data from VectorBase or other entity

    Example: "We have also implemented *data mining* tools, such as the HMMER package (http://hmmer.janeliex.org/) to build profile hidden Markov models from multiple sequence alignments which can then be used for sensitive database *searching* using statistical descriptions of a sequence families consensus"
14. **Manual/automatic:** Any explicit mention of a process being manual or automated (i.e., performed by human vs. computer)

    Example: "The annotation of the An. gambiae genome is being *manually* appraised using the GMOD annotation tool Apollo (4). Currently, over 50 % of the genome has been completed including the entirety of the chromosome arms 2L, 2R and X."
15. **Future work:** Any mention of work to be done (construed in the future tense), or mention of future directions or goals

    Example: "A number of Anopheles species *will be targeted* for genome sequencing (http://www.vectorbase.org/Docs/ShowDoc/?doc=White Papers) and the reduction in cost means that individual labs can produce significant amounts of sequence data from species or isolates. The integration and management of these data will be a major challenge for *the coming years.*"
16. **Community:** Any mention of users, a community of scientists, work being outsourced to a community, and so on; often co-occurs with "collaboration"

    Example: "Other material of interest to the vector community is being incorporated, including the newly developed controlled vocabulary of mosquito anatomy (http://obo.sourceforge.net/detail.cgi?mosquito_anatomy) and other vector-related ontologies."
17. **Customization:** Any mention of the database interface being tailored for/by a specific user

    Example: "The standard display methods provide a wide variety of options that can be customized by a submitter to best suit their datex."

18. **Reference to past report:** Any explicit mention to a previously published VectorBase report; often co-occurs with "past growth"

    Example: "As anticipated in *our previous update* (2), analyses of populations and variations at the genomic level have increased significantly."

19. **Past growth:** Any mention of improvement to the database construed in the past tense

    Example: "A major overhaul of the PopBio infrastructure *was undertaken in 2012–13* and a few of the more user-visible changes are described here. URL robustness and data maintainability *were improved* by the allocation of stable IDs for projects, samples and assays. The submission procedure *was streamlined* to a single ISA-Tab (9) spreadsheet submission."

20. **Breadth/scope of data:** Any mention or *display* of the breadth, scope, or variety of data included in VectorBase

    Example: "VectorBase currently hosts *nine genomes* of which the majority are mosquitoes, reflecting their importance in disease agent transmission. The seven corresponding species are: *Anopheles gambiae (three genomes, for the PEST, Mali-NIH and Pimperena colonies), Aedes aegypti, Culex quinquefasciatus, Glossina morsitans, Ixodes scapularis, Pediculus humanus and Rhodnius prolixus.*"

Notes

1. Dennis A. Benson et al., "GenBank," *Nucleic Acids Research* 41, no. D1 (2013).
2. Sabina Leonelli & Rachel A. Ankeny, "Re-thinking Organisms: The Impact of Databases on Model Organism Biology" *Studies in History and Philosophy of Biological and Biomedical Sciences* 43, no. 1 (2012).
3. Carolyn R. Miller, "The Aristotelian Topos: Hunting for Novelty" in *Rereading Aristotle's Rhetoric*, ed. Alan G. Gross & Arthur E. Walzer (Carbondale: Southern Illinois University Press, 2000).
4. Bruno J Strasser, "The Experimenter's Museum: GenBank, Natural History, and the Moral Economies of Biomedicine," *Isis* 102, no. 1 (2011).
5. Jeanne Fahnestock and Marie Secor, "The Stases in Scientific and Literary Argument," *Written Communication* 5, no. 4 (1988).
6. Ibid.
7. Daniel Lawson et al., "VectorBase: A Home for Invertebrate Vectors of Human Pathogens," *Nucleic Acids Research* 35, Database issue (2007): D503.

8. Ibid.
9. Daniel Lawson et al., "VectorBase: A Data Resource for Invertebrate Vector Genomics," *Nucleic Acids Research* 37, Database issue (2009).
10. Karine Megy et al., "VectorBase: Improvements to a Bioinformatics Resource for Invertebrate Vector Genomics," *Nucleic Acids Research* 40, Database issue (2012).
11. G. I. Giraldo-Calderon et al., "VectorBase: An Updated Bioinformatics Resource for Invertebrate Vectors and Other Organisms Related with Human Diseases," *Nucleic Acids Research* 43, no. Database issue (2015).
12. Michael Y Galperin, Daniel J. Rigden, and Xose M. Fernandez-Suarez, "The 2015 Nucleic Acids Research Database Issue and Molecular Biology Database Collection," *Nucleic Acids Research* 43, no. Database Issue (2015): D1.
13. Lynda Walsh, "Resistance and Common Ground as Functions of Mis/aligned Attitudes: A Filter-Theory Analysis of Ranchers' Writings About the Mexican Wolf Blue Range Reintroduction Project," *Written Communication* 30, no. 4 (2013).
14. Elvira Mitraka et al., "Describing the Breakbone Fever: IDODEN, an Ontology for Dengue Fever," *PLOS Neglected Tropical Diseases* 9, no. 2 (2015).
15. David Ribes and Geoffrey C. Bowker, "Between Meaning and Machine: Learning to Represent the Knowledge of Communities," *Information and Organization* 19, no. 4 (2009): 199.
16. Susan Leigh Star and James R Griesemer, "Institutional Ecology, 'Translations' and Boundary Objects: Amateurs and Professionals in Berkeley's Museum of Vertebrate Zoology, 1907–39," *Social Studies of Science* 19, no. 3 (1989).
17. Ribes and Bowker, "Between Meaning and Machine."
18. Ibid., 214.
19. Pantelis Topalis et al., "IDOMAL: An Ontology for Malaria.," *Malaria Journal* 9, no. 230 (January 2010): 1.
20. Ibid., 2.
21. Ribes and Bowker, "Between Meaning and Machine."
22. Ibid., 215.
23. Topalis et al., "IDOMAL," 2.
24. Ibid., 8.
25. Mitraka et al., "Describing the Breakbone Fever," 2.
26. Carolyn R. Miller and Jack Selzer, (1985). "Special Topics of Argument in Engineering Reports," in *Writing in Non-Academic Settings,* ed. Lee Odell & Dixie Goswami (New York: The Guilford Press, 1985).
27. Lawrence J. Prelli, *A Rhetoric of Science: Inventing Scientific Discourse* (Columbia, SC: University of South Carolina Press, 1989).

28. Lynda Walsh, "The Common Topoi of STEM Discourse: An Apologia and Methodological Proposal, With Pilot Survey," *Written Communication* 27, no. 1 (2010).
29. Ibid.
30. Prelli, *A Rhetoric of Science.*
31. Mitraka et al. "Describing the Breakbone Fever."
32. Walsh, "The Common Topoi of STEM Discourse."
33. Ibid., 128.
34. Star and Griesemer, "Institutional Ecology, 'Translations' and Boundary Objects."
35. Michael Leff, "Topical Invention and Metaphoric Interaction," *The Southern Speech Communication Journal* 48, no. 3 (1983).
36. Prelli, *A Rhetoric of Science*, 66.

Bibliography

Beatty, John. "Speaking of Species: Darwin's Strategy." In *The Units of Evolution*, ed. Michael Ereshefsky, 227–246. Cambridge: MIT Press.

Benson, Dennis A., Mark Cavanaugh, Karen Clark, Ilene Karsch-Mizrachi, David J. Lipman, James Ostell, and Eric W. Sayers. 2013. "GenBank." *Nucleic Acids Research* 41, no. D1: 36–42.

Fahnestock, Jeanne, and Marie Secor. 1988. "The Stases in Scientific and Literary Argument." *Written Communication* 5, no. 4: 427–443.

Galperin, Michael Y., Daniel J. Rigden, and Xose M. Fernandez-Suarez. 2015. "The 2015 Nucleic Acids Research Database Issue and Molecular Biology Database Collection." *Nucleic Acids Research* 43, Database issue: D1–D5.

Giraldo-Calderon, G.I., S.J. Emrich, R.M. MacCallum, G. Maslen, E. Dialynas, P. Topalis, N. Ho, et al. 2015. "VectorBase: An Updated Bioinformatics Resource for Invertebrate Vectors and Other Organisms Related with Human Diseases." *Nucleic Acids Research* 43, Database issue: D707–D713.

Lawson, Daniel, Peter Arensburger, Peter Atkinson, Nora J. Besansky, Robert V. Bruggner, Ryan Butler, Kathryn S. Campbell, et al. 2007. "VectorBase: A Home for Invertebrate Vectors of Human Pathogens." *Nucleic Acids Research* 35, Database issue: D503–D505.

———. 2009. "VectorBase: A Data Resource for Invertebrate Vector Genomics." *Nucleic Acids Research* 37, Database issue: 583–587.

Leff, Michael. 1983. "Topical Invention and Metaphoric Interaction." *The Southern Speech Communication Journal* 48, no. 3: 214–229.

Leonelli, Sabina, and Rachel Ankeny. 2012. "Re-thinking Organisms: The Impact of Databases on Model Organism Biology." *Studies in History and Philosophy of Biological and Biomedical Sciences* 43, no. 1: 29–36. Elsevier Ltd

Megy, Karine, Scott J. Emrich, Daniel Lawson, David Campbell, Emmanuel Dialynas, Daniel S.T. Hughes, Gautier Koscielny, et al. 2012. "VectorBase: Improvements to a Bioinformatics Resource for Invertebrate Vector Genomics." *Nucleic Acids Research* 40, Database issue: D729–D734.

Miller, Carolyn R. 2000. "The Aristotelian Topos: Hunting for Novelty." In *Rereading Aristotle's Rhetoric*, ed. Alan G. Gross and Arthur E. Walzer, 130–146. Carbondale: Southern Illinois University Press.

Miller, Carolyn R., and Jack Selzer. 1985. "Special Topics of Argument in Engineering Reports." In *Writing in Non-Academic Settings*, ed. Odell Lee and Dixie Goswami, 309–341. New York: The Guilford Press.

Mitraka, Elvira, Pantelis Topalis, Vicky Dritsou, Emmanuel Dialynas, and Christos Louis. 2015. "Describing the Breakbone Fever: IDODEN, an Ontology for Dengue Fever." *PLOS Neglected Tropical Diseases* 9, no. 2: 1–19.

Prelli, Lawrence J. 1989. *A Rhetoric of Science: Inventing Scientific Discourse*. Columbia: University of South Carolina Press.

Ribes, David, and Geoffrey C. Bowker. 2009. "Between Meaning and Machine: Learning to Represent the Knowledge of Communities." *Information and Organization* 19, no. 4: 199–217.

Star, Susan Leigh, and James R. Griesemer. 1989. "Institutional Ecology, 'Translations' and Boundary Objects: Amateurs and Professionals in Berkeley's Museum of Vertebrate Zoology, 1907–39." *Social Studies of Science* 19, no. 3: 387–420.

Strasser, Bruno J. 2011. "The Experimenter's Museum: GenBank, Natural History, and the Moral Economies of Biomedicine." *Isis* 102, no. 1: 60–96.

Topalis, Pantelis, Elvira Mitraka, Ioana Bujila, Elena Deligianni, Emmanuel Dialynas, Inga Siden-Kiamos, Marita Troye-Blomberg, and Christos Louis. 2010. "IDOMAL: An Ontology for Malaria." *Malaria Journal* 9, no. 230: 1–11.

Walsh, Lynda. 2010. "The Common Topoi of STEM Discourse: An Apologia and Methodological Proposal, With Pilot Survey." *Written Communication* 27, no. 1: 120–156.

———. 2013. "Resistance and Common Ground as Functions of Mis/Aligned Attitudes: A Filter-Theory Analysis of Ranchers' Writings About the Mexican Wolf Blue Range Reintroduction Project." *Written Communication* 30, no. 4: 458–487.

CHAPTER 6

Genre Signals in Textual Topologies

William Hart-Davidson and Ryan Omizo

North American genre theory, also known as rhetorical genre studies[1] provides several powerful and counterintuitive conjectures that have thus far proven difficult to explore systematically, let alone "test," due to the scale of analysis required. One, raised by Schryer,[2] is the idea that textual regularities that we come to recognize as the formal features of a genre are only relatively stable across instances, and that these stabilities fluctuate over time rendering them only "stable for now" according to Schryer.[3] Following from this idea, any one instance of a genre recognized as such should exhibit formal qualities at some detectable level of regularity, but we could also expect significant differences between one instance and any other, and we could expect to trace "drift" over time in the same nominal genre category.

Using qualitative methods, researchers such as Spinuzzi,[4] Teston et al.,[5] and others have based studies on these basic concepts in genre theory, producing evidence that supports the logic of what Miller[6] first characterized

W. Hart-Davidson (✉)
Department of Writing, Rhetoric, and American Cultures, Michigan State University, East Lansing, MI, USA

R. Omizo
Harrington School of Communication and Media, University of Rhode Island, Kingston, RI, USA

© The Author(s) 2017

L. Walsh, C. Boyle (eds.), *Topologies as Techniques for a Post-Critical Rhetoric*, DOI 10.1007/978-3-319-51268-6_6

as "genre as social action." But few researchers, apart from Miller,[7] have investigated the core ideas of genre theory as "basic science." One reason may simply be that the methods to explore questions such as "if genre stability depends on repetition in response to recurring exigencies (leading to stability), but not repetition at the syntactic level (accounting for variation across instances), what *are the structures that are repeated*?" have not been readily available.

In this chapter, we propose two topological transformations that enable textual analysis at a scale sufficient to investigate some of the central propositions of genre theory. We use the term *topology* as a means to focus our transformation and analysis on the mathematical relationships between elements such as nodes and links in a network as opposed to *topography*, which suggests a thick description of physical elements as in a terrain map. We outline both topological techniques from their roots in theories of written discourse to a stepwise procedure for turning text into a computational object suitable to topographic analysis: a graph.[8] Most forms of analysis, but particularly analysis at scale, require researchers to transform texts to be processed algorithmically—by means of a systematic and repeatable procedure, whether executed by humans or machines. For this reason, we take care to discuss the ways our topographical transformations retain key features of their natural language text counterparts in support of rhetorical analysis.

The first transformation renders the words of a text as a topology that may be analyzed following the principles of graph theory. How a given text is built up and where it conforms to or diverges from an expected structural sequence are revealed with this transformation. We highlight one straightforward example: comparison of texts by the similarity of their corresponding adjacency matrices. With words transformed into nodes and the relationships among these nodes recorded as edges in the graph, we can detect similarities of structure that do not depend on exact words being the same. On the contrary, substitutions are quite well tolerated with rhetorical structures intact, just as they are when readers encounter actual texts.

The second transformation constructs network graphs of discourse. This approach allows us to map recurring moves within and across multiple text sequences as traversals of a network graph structure. With this technique, we can identify where key topics become *commonplaces* (*topoi*) that stabilize and focus discourse. One key stabilizing action is the repetition of familiar discursive structures. In these instances, writers send genre signals and construct paths that return to "commonplaces" in order to keep the social contract of shared expectations, an idea inherent in the theory of genre as social action. Ultimately, this chapter presents a theoretical exploration that undergirds work we have done to create software that performs rhetorical analyses in a manner designed to be assistive to rather than replacing human interpretation. In this work, we have taken an explicitly topological approach to text and discourse analysis. We understand our topological approach to analysis to work within an acceptable range of accuracy and, most importantly, usefulness. We have published this work, including usable versions of the software, in other venues that we invite readers to explore.[9] Our purpose here is not to explicate the computational methods used in this prior work, however. The work of this chapter is to explore how and why a topological approach may be effective and useful for rhetorical analysis. We also aim to contribute to an understanding of the limits of topological methods for producing rhetorical technology that might augment human interpretation. In presenting these techniques, we seek to join researchers such as Graham *et al.*,[10] in providing fellow rhetorical scholars with a foundation for applying computational methods of analysis to our work that are consistent with the theoretical tools that inspire the questions we may seek to answer.

Topology and Text Transformation

There are two important meanings of topology that we rely on to make useful contributions to rhetoric theory via our work on textual transformations as seen in genre theory. Both of these meanings are present in the basic definition of topology as a mathematical principle. The first speaks of topology as a way to understand and measure spatial relationships among elements. The second meaning is a bit more obscure and harder to grasp without knowing something about the origins of topology in mathematics.[11] This is the sense that topology is fundamentally about the connective properties that constitute an object or set of objects in space and that remain true or observable under the conditions of some form of continuous deformation such as stretching or bending.

If we limited ourselves to just the first meaning of topology, it might seem to be a problem to suggest that we can usefully represent a text as a topology for the purposes of representing its rhetorical meaning. Seeing a group of texts this way is sometimes useful—for example, when we depict the structure of a hypertext—but how much rhetorical meaning does a display like this convey? Not very much. But when we add the second meaning to the first, the value of topological transformations starts to become more clear for the rhetorician. A topological approach gives us a set of analytical tools to understand how we can bend and stretch rhetorical figures, even substitute or leave out words[12] and still detect rhetorical moves, tropes, or figures that are recognizable as instances of a common type.

Here, we wish to point out that we are not merely talking about using the analytical methods of topology metaphorically for rhetorical analysis. In our computational work, we are using these methods literally as both an analytic to uncover insights and as means of inventing new parameters for inquiry. To do that, however, we need to represent a text as a topological structure: a graph. In the next section, we demonstrate this conversion. The choices we make during transformation impact the kinds of rhetorical analysis we can perform, and we make different choices in different situations to construct a text as a graph. But before we discuss these transformations specifically, we explore the idea of texts as topologies more generally and ask: why does this idea of seeing a text as a graph make sense?

Rhetoricians already talk about texts in topological terms. Texts are compared to networks because of their distributed and interconnected nature, although this distribution and relationality is not always mathematically quantified. One concept for which a topological viewpoint has proven to be especially advantageous is the notion of genre. This makes sense given that the study of genre is the study of certain kinds of textual stabilities, of persistent structures, in the context of language use. In both synchronic and diachronic studies of genre, we can similarly understand the desire to reconcile genre variations as an analysis of relatively stable structures in the midst of constant variation. Thus, it is not surprising to find descriptions like the following in our foundational theoretical works on genre from Bakhtin:

> the utterance proves to be a very complex and multiplanar phenomenon if considered not in isolation and with respect to its author (the speaker) only, but as a link in the chain of speech communication with respect to other, related utterances.[13]

This passage follows a much more well-known passage elaborating on his famous idea of "heteroglossia," the notion that to be ***understood*** by others we build utterances from words ***used*** by others. The conception of an utterance as consisting of links in a chain is explicitly topological. The idea that in order to make meaning we must construct utterances in precisely this way is also consistent with the idea of *topoi* as commonplaces. That is, we must build utterances as through points we and others have visited before. Bakhtin's essay on speech genres offers perhaps the most well-known elaboration of this idea that we can understand language use as a concrete set of moves, which enable us to act with others by putting words in patterns meant to be taken up and compared with previous patterns of similar signals.

Paré[14] represents the North American genre theory perspective when writing about how the concrete actions of academic work give rise to textual stabilities over time. He writes:

> [a]ny given text emerges from a sequence of more-or-less predictable actions (literature reviews, note-taking, data collection, drafting, peer review, etc.), comes to life within the customary forums and practices of the academy (courses, exams, conferences, journals), and has consequences within the constraints of disciplinary and institutional cultures (grades, credentials, tenure, etc.).[15]

Paré's example explicates the phenomenon of textual stability that arises due to Miller's[16] definition of genre as social action. Habitual responses to recurring situations produce stable discursive structures. But Paré's example also accounts for variation, for the stretching and deformation that can occur due to the choices individuals make when responding to similar situations which are never identical and which only occasion but never fully prescribe a specific response. Schryer[17] captures this idea in her description of generic structures as "stable for now," allowing for the observable patterns of change in utterances as exigencies change.

In textual topologies, generic utterances deform across time. If we imagine a mathematician using topological methods to study objects that stretch and bend in space, we as rhetoricians use them to study the way genres mutate over time. By doing so, we can learn which structures are robust, how relationships among elements change as they undergo change, and where ruptures may occur. We can do this, moreover, by tuning into the same signals that interlocutors or writers and readers do. As analysts, we should expect to detect some strong signals corresponding to commonplaces, those stable-for-now moments of language reuse that we utter to make ourselves understood. We should also expect weaker signals, moments in specific rhetorical situations wherein we make our utterances novel by performing what Derrida calls *différance*.[18]

To gain the affordances of topological analysis, we have to render a text or a set of texts spatially. This usually means that we must make the text into a different kind of *computational object* than what appears on a sheet of paper or on a screen. By computational object we mean something that you can count. These counts can be as simple as word frequencies or something as sophisticated as weighted term document matrix that accounts for the presence or absence of terms in corpora. The methods we discuss in the remainder of this chapter transform texts into topologies for the purpose of conducting rhetorical analysis. We are not the first or only researchers to transform texts into graphs or suggest that it may be helpful to perform topological analyses of texts to reveal patterns.[19] We show how such transformations can be done and what analytic possibilities follow from these transformations for scholars who are looking to understand recurring language structures as they are used, consistent with Paré's description above, in specific kinds of social contexts, subject to particular kinds of social, cultural, or institutional influences.

Transforming Texts to Graphs

Following from the ideas by Miller and Schryer, Paré[20] notes that across relatively similar situations that give rise to textual regularity there are nonetheless specific conditions to which a writer must respond. Texts are, in Paré's words, "always custom." This poses an interesting and thorny problem for rhetorical learners—the human and the machine-learning algorithmic kind. The problem is that where there are bits that are reused or repeated to make a signal aligned with a particular genre, what is repeated or reused goes beyond mere words. When the occasions and the interlocutors vary, the precise words and the order of those may vary as well. This means that while there might be repeated signals, certain kinds of important rhetorical moves are not inherent in the meaning of the words.

An example of the way nonhuman learners can get tripped up by this quality of texts being "always custom" as Paré neatly puts it, uses a tool made available by the Advanced Marketing Institute named "Headline Analyzer." The tool is promoted as a resource to help users craft clickworthy headlines for online content, the main property of which is that they contain what the AMI calls "emotional marketing value." The user can paste a headline into the tool and the headline analyzer gives the text a score to indicate if it has sufficient "emotional marketing value."

To test the tool's performance, we supplied the following headline from the social media site *Upworthy* published on March 5, 2014, by Joseph Lamour: "His Classmates Thought Making Fun of Him Made Perfect Sense But Then a Senseless Thing Happened." *Upworthy* is a viral news site that uses a method called A/B testing to discover the most effective headlines, which it measures by counting the actual number of readers to click through and read one version of a headline vs. other versions. Multiple versions of headlines for the same stories are tested against one another. Those headlines with the most clicks are deemed the "winner." The headline noted above was one of these superior performers. However, when we selected one of this heartstring winner and ran it through the Headline Analyzer, we received unexpected results. The output appears in Fig. 6.1.

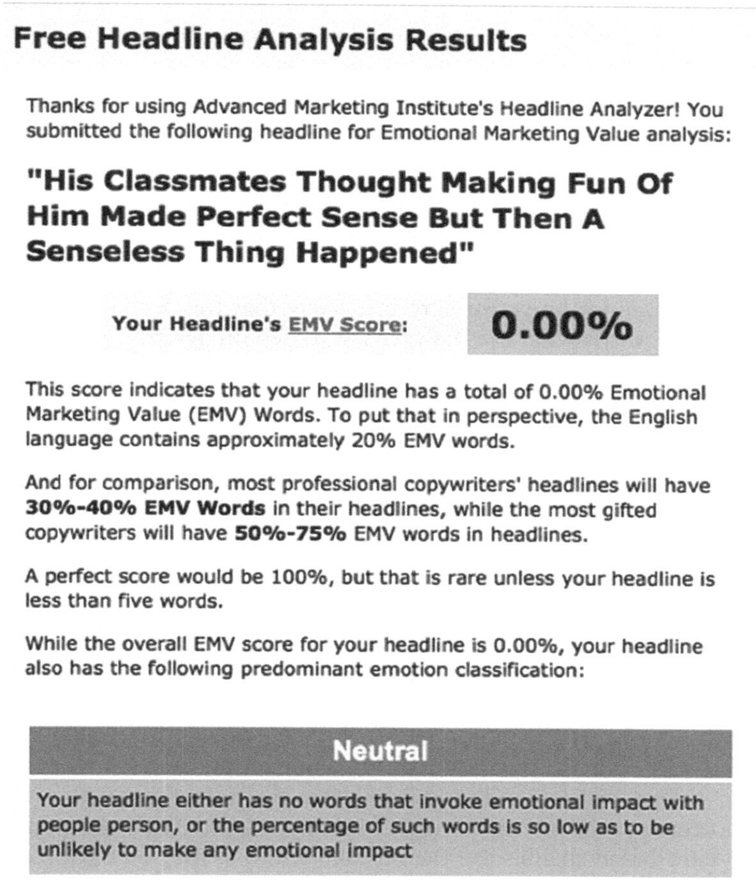

Fig. 6.1 Headline analyzer results

The results indicate that the *Upworthy* headline has no emotional impact. The readers of *Upworthy*, however, may not necessarily agree. Neither do we. What is this analysis tool missing that we see? Rhetorically speaking it is an appeal to pathos. But if we examine the way this particular headline analyzer works, we can start to see why the results may be skewing. Based on the description of the results, we understand that this analyzer looks at emotional words and assigns the text a score based on the presence of words that are thought to have emotional content. It seems to perform other calculations too, such as the frequency of emotionally charged terms or a calculation of the ratio of emotional words to nonemotional words, though we cannot be sure exactly how based on this description. What we can say, however, is that just because there are no emotional word matches in this particular headline does not mean there are no emotional appeals in this headline. Here is the headline again:

> **His Classmates Thought Making Fun of Him Made Perfect Sense But Then a Senseless Thing Happened**

Though we cannot know for sure it is likely that the headline as shown above is not exactly what the analyzer saw. Based on descriptions of similar kinds of analysis,[21] we speculate that Headline Analyzer analyzed the following vocabulary:

> classmate think make fun made perfect sense thing happen

This is a list of word "tokens" with some of the most common types of words such as articles and prepositions removed and other words transformed slightly to represent their dictionary roots or stems. The verb *thought* is represented above, for example, as its present tense imperative form "think." This kind of processing is done to both boost the "signal" and reduce what is thought to be "noise" in an analysis like this one. Stemming words boost the signal of repetition by allowing the computer to see that forms of a common root word are actually the same semantic token inflected for grammatical reasons. Eliminating function words such as prepositions, articles, and pronouns cleans the text of words thought to have no emotional content, leaving a less noisy signal. Without this kind of step, a ratio-based analysis like the one alluded to in the description of the results of the headline analyzer could be unduly influenced by the presence of words such as "him" or "a." After these (likely) text processing steps are performed, the resulting set of tokens was judged by the headline analyzer to contain no emotional value. Of the tokens that remain, only one seems to be strongly tied to emotion: fun. The others are, as the analyzer suggests, relatively neutral.

In order to see how we might get a different kind of result from an analysis designed to reveal pathos, one that aligns more with the way human readers understand and recognize an emotional appeal in this short headline, we could turn the text into a graph instead of a normalized list of words. The headline is represented as a series of points and lines, nodes and edges in Fig. 6.2.

We also performed some text processing to boost signal and dampen noise. Here is our token list:

classmate think make fun him perfect but then sense thing happen

The graph uses two textual features of the token list for its construction: adjacency and repetition. The rationale for this follows the logic of Bakhtin. We build the graph as the utterance flows, each token becomes a node connected to the next one with an edge. When a token repeats, we link back to the first instance of that node, and we see additional edges start to form. The result is something called a directed graph that follows the sequence of the utterance. Two tokens in this example are repeated: make and sense. In the image of the graph in Fig. 6.2, we have indicated differences in our stop-words with the red nodes. These are words that typically are deleted but remain in our version: "him," "but," "then."

Fig. 6.2 Network graph of headline

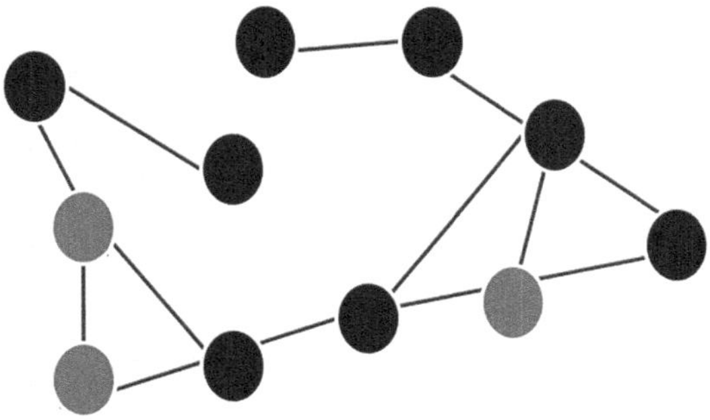

Revisiting the original text, there is a rhythm to the sentence that is established in part by the repetition of words that are set up as opposites: sense and senseless. Rhetorically speaking, the headline is an example of chiasmus (itself a topological figure that traverses similar clauses to make a point). Our graph captures that structure with links. The two gray tokens 'but' and 'then' also signal a narrative reversal, a kind of dramatic convention that allows the reader of the headline to get a sense of what the story will be by hinting at a narrative structure that is familiar: the twist at the end. This symmetry is more obvious if we represent the graph using an adjacency matrix. In mathematical terms Fig. 6.3 is the same as Fig. 6.2.

Each line along the x and y axes in the matrix in Fig. 6.3 represents a node from the tokenized *Upworthy* headline. It has a one or a zero corresponding to indicate the presence or absence of a link with every other node in the graph. There are 11 nodes in the graph, so this is an 11 by 11 matrix. The first node, represented by the first horizontal line, has links to just one other node. This relationship is represented by two ones in position one and two in line one of the matrix. We thus have a global representation of one aspect of the structure of the graph in a simple format.

The matrix serves as a fingerprint for the headline text, one that can be compared with other fingerprints. And because we are no longer looking at

Fig. 6.3 Adjacency matrix of *Upworthy* headline

1	1	0	0	0	0	0	0	0	0	0
1	1	1	0	0	0	0	0	0	0	0
0	1	1	1	1	1	0	0	0	0	0
0	0	0	1	1	1	0	0	0	0	0
0	0	1	1	1	1	0	0	0	0	0
0	0	1	0	1	1	1	0	0	0	0
0	0	0	0	0	1	1	1	1	0	0
0	0	0	0	0	0	1	1	1	0	0
0	0	0	0	0	0	1	1	1	1	0
0	0	0	0	0	0	0	0	1	1	1
0	0	0	0	0	0	0	0	0	1	1

words, these comparisons are tolerant of substitutions and subtle rearrangements, but also sensitive to the kind of narrative structural cues that may communicate a pathetic appeal. In the viral headline example, we made decisions to transform the text into a graph in a way that made use of two specific features we thought might help to capture the structure we were looking for. We cannot know for sure if it does by looking at a single example. But what we have done is provide a consistent method to compare many examples, and we have given ourselves a wide range of options for evaluating the resulting structures. We close this section by reviewing a key network graph structure: betweenness centrality. Figure 6.4 shows a simple graph arranged and labeled to highlight some features of each node as well as the overall graph structure.

The letters label each node and the number in parenthesis indicates the degree of each node. At a glance, we can see that node D has the most connections. It is connected to every other node in the graph. Moreover, node D occupies a central position in the overall structure relative to the other nodes. In graph theory terminology, we say node D has high betweenness centrality. Centrality is measured by examining the shortest path from any one node to other nodes. Nodes B and E are less central. Knowing something about the betweenness centrality of a node can be useful. For example, if the graph in Fig. 6.4 represents a telephone network and node D ceases broadcasting, much of the neighborhood would be affected. Node D occupies a "brokerage position" in the graph, meaning that a message traveling from the left side to the right side of the network would need to pass through node D to get to its destination. By contrast, if something happened to node B, the effects to the rest of the network would likely be minimal because of its low centrality. Measures of centrality are key parts of well-known algorithms such as Google's PageRank, a measure used to compute the relevance of search results.[22] Centrality measures can also be used to find important topics in a text or text corpus.[23]

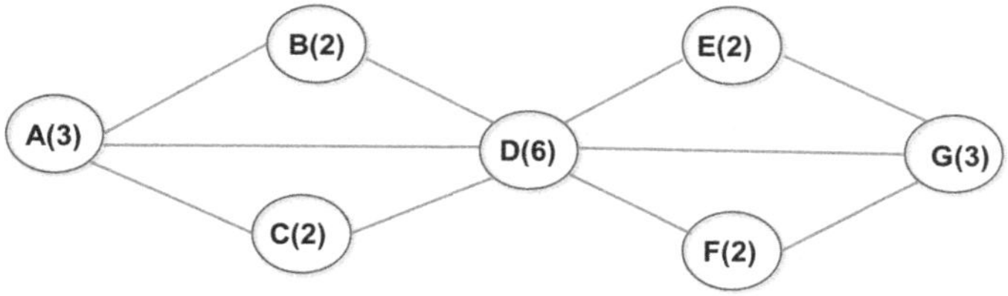

Fig. 6.4 A fully connected network graph

We want to stress that this is not the only way to make a text into a graph, nor is it necessarily the best way. It is a deliberately simplistic way that we propose here to show how we might gain the affordances of graph analysis in the context of a specific rhetorical analysis problem. In this example, we were guided in our choices by a desire to capture what was both stable and "custom" in a viral headline. We were also working with a relatively small and simple text. In the next section, we will address challenges of working with larger and large numbers of texts, situations that call not only for different approaches to transforming texts into graphs but also open up new possibilities for analyzing the results.

Discourses and Graph Traversals

The fundamental process of analyzing a topology rather famously involves working out pathways among the vertices. The origin story of topology as an area of mathematical study, in fact, begins with the mathematician Leonhard Euler demonstrating that a problem known as The Seven Bridges of Königsberg—a puzzle to find a nonredundant path across the Russian city of that name—was not resolvable. We can understand how working out the shortest or the least redundant traversal through a number of points can be helpful. We can imagine the benefits to a mail carrier. But we understand traversals as having an important correspondence to rhetorical analysis as well. However, unlike the delivery service that wants to eliminate paths with repeat visits to a node in the interest of efficiency, we find that as rhetoricians we often want to focus on repeat visits and moments of return when charting a traversal. The reason is tied to the notion of genre signals we introduced earlier, as well as to the rhetorical concept of topics or commonplaces. As Miller[24] writes, commonplaces make connections between ideas and idea types (e.g., concrete to the abstract), and facilitate the recognizability of these ideas by offering the reader familiar paths in the interpretive process. In discourse, whether carried out orally or through an exchange of texts, we understand the need to return to common points of interest as a fundamental means of staying "on topic" or adhering to a particular group's conventions for interacting. This is often how we make ourselves (a) understood, and (b) recognized as being a valid contributor to a particular discourse community. We even have common ways of talking about problems that echo the idea of traversals. We talk about "wandering off" a topic when we want to characterize a discourse as incoherent, for example. We might also talk about a conversation "going in circles" when it may visit the same topics too many times. An interlocutor that seems to be ignoring conventions is similarly said to be on the wrong path or "out in left field"—a bit of a mixed metaphor in this case that nonetheless speaks to the property of centrality or connectedness that we discussed earlier.

When we transform discourse into a graph, we render it as a surface to map. And when we trace paths across the surface of a textual topology, we can see commonplaces as common (or not very common) by virtue of how many times they are visited in a given traversal. If we add a temporal dimension to our analysis, we gain the ability to add frequency as a measure of what is "common" or not in a given discourse.

For example, in one of our recent projects, we examined discussion threads from an online forum, a blog about science and technology topics.[25] Our aim in this project was to develop a means by which we might characterize the discussion in a set of comments as on topic or not. In most cases with the threads we were examining, "on topic" meant that participants were talking about a topic introduced in the main post *and* doing so in a way that bore some resemblance to scientific discourse. To be sure, we were not expecting a comment section on the Internet to be similar to scientists writing in peer-reviewed journals. We expected to see an informal discussion. But we were interested in learning whether we might see some similarities between the way scientists write and the way people talking about a topic write when they are talking scientifically. We also hoped, in the end, to be able train an algorithm to make some distinctions between threads that treated a topic in more and less scientific ways, even when they were talking about the same topic.

Our primary objects of inquiry for this study were a set of discussion threads that had already been read and evaluated by a team of fellow researchers, so we could begin our work by examining what our colleagues had deemed to be successful threads. In these threads, participants were engaged in a discussion of a relevant topic and were carrying out the discussion in a manner that demonstrated some evidence of scientific reasoning, possibly even learning, was going on over time. Of these successful threads, one stood among the rest for its durability and robust participation: a discussion of raising chickens in urban settings. When we first began working with it, the chicken thread had been going strong for over seven years and had hundreds and hundreds of comment posts.

To begin our exploratory analysis, we performed the same kinds of processing steps we discussed above to boost signal and dampen noise, and then we produced a very simple graph using word adjacencies and repetition to create nodes and edges. The image in Fig. 6.5 is what we saw.

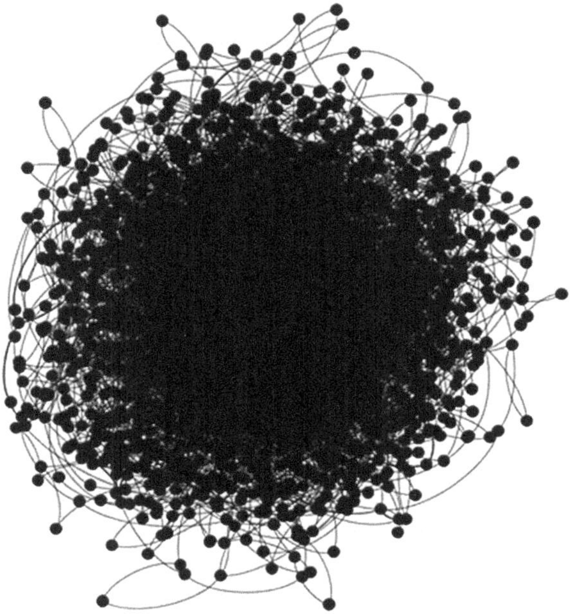

Fig. 6.5 Network graph of "chicken thread"

This large, dense graph presented in Fig. 6.5 is obviously not very valuable for processing visually other than to give a sense of just how large the text we were working with was. It is what network analysts call a "hairball." But because it is a computational object, we could perform some analysis of it in this form. With the help of a computer, we calculated the centrality of the vertices in the graph to see which topics were common. The highest betweenness centrality tokens from that analysis are shown in Table 6.1.

A few interesting insights emerge from this analysis. Firstly, the topic of the thread is quite clear and statistically significant. We can see the chicken topic in words like "egg," "chicken," "hen," "rooster,", and "chick." Note, too, that most of these words are even more prevalent than what in this sample are noise tokens—"posted" and "2007"—tokens that appear not because the participants type them but as parts of an automated timestamp that accompanies each message. The high betweenness centrality of these terms in our graph marks their reiteration over time within and across comments to the thread. They are commonplaces to which the participants return, time and time again, over seven years.

Table 6.1 High betweenness centrality terms in the "chicken thread"

Token ID	*Betweenness centrality*
egg	0.338034594
if	0.102616759
hen	0.096716837
chicken	0.094978886
say	0.089044754
not	0.07245774
can	0.067463635
will	0.04860393
rooster	0.047524128
day	0.046672289
10	0.043184336
2007	0.04213273
chick	0.040569829
posted	0.040274341

We constructed this network as a directed multiedged graph. This means that the edges have a direction—pointing to the adjacent node next in the sequence. Moreover, a single node such as "egg" could have multiple edges running in and out of it, which explains why multiple loops appear between similar nodes. Our graph included directed edges to represent the connections between words as they appear in sequence in the text moving from left to right as we conventionally read in English. When we zoom in to reveal the structure of the large graph in finer detail, we can see how the centrality of the commonplace terms forms looping structures in the graph. In Fig. 6.6 the large arrow shows how egg is a central node in several of these cycles. The large "hairball" is composed of many of these loops.

What does an "on topic" thread look like? One answer we can give from this analysis is that not very many tokens go by before we see one of the high betweenness centrality topic words like "egg" or "chicken" recur (see Fig. 6.6). In graph theory terms, these kinds of looping structures are sometimes called walks. We can gain a sense of the structure of discourse is to look at how long the walks are that return to an important topic word. This helps us to see how tight the loops are, how frequently the topics are reiterated, and thus how focused the discourse is on a particular topic. In the chicken thread, we noticed a remarkable pattern that not only gave us

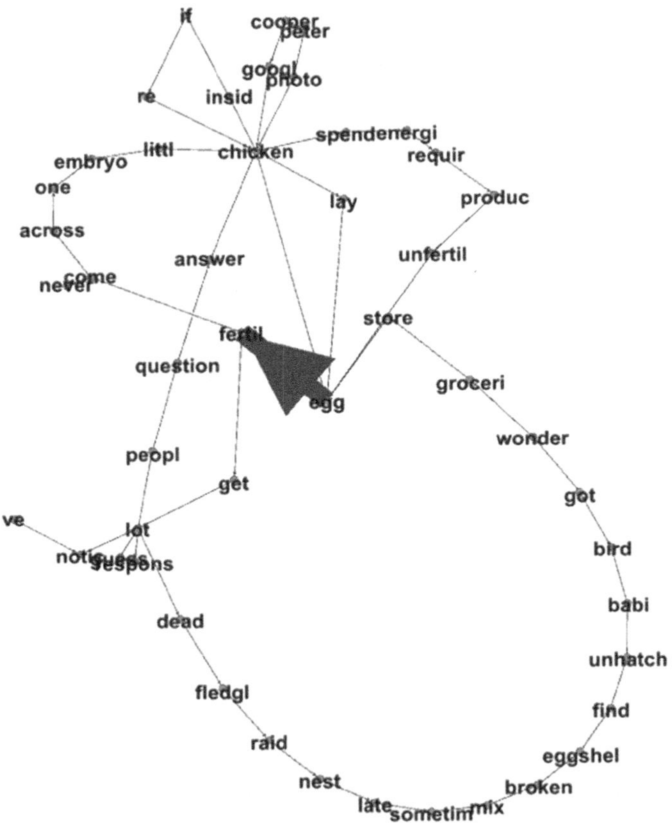

Fig. 6.6 Zoomed in view of the "chicken thread" network graph

an indication of what the participants were talking about, but how they were talking about chickens and eggs.

When we looked at the loops in the network graph, a pattern emerged. This was exciting to us because we could compare these passages with those in the natural language text where the human raters had seen evidence of stable communicative patterns that involve the use of conditional questions and statements framed around the term "if." Additionally, modal verbs pervaded the pattern, which would be expected based on the syntactic use of *if* in English. As a conditional, "if" can signal subjunctive reasoning.

For example, "*if* x, then y *can* occur" or "*if* x, then y *will* occur." Table 6.1 provides additional support. Among the high centrality topic words are tokens that are used to hedge claims: "if," "when," "can," and "will." These modal verbs and conditional terms help us see that participants are not only talking about eggs and chickens, but that they are doing it in a way that suggests they are sensitive to the relationship between claims and evidence and are using qualifiers to aid precision, survey for multiple aspects of an issue, and account for previous claims by linking the 'if' clause with an earlier premise.

One other outcome of these exploratory analyses is a re-imagining of a network graph visualization that separates the tangles routinely created by large network graphs. We have termed this prototype visualization the "Walk-a-Tron." This visualization presents the repeated occurrences of a high betweenness centrality term as adjacent circles. These circles, in part, signal the repetitive or cyclic aspects of these high betweenness centrality terms. At the same time, the radii of these circles are given by the number of word-nodes that are interposed between the next occurrence of the high betweenness centrality term. Put another way, the radii of each circle accounts for the sum of intervening edges between n and $n + 1$ occurrences of a high betweenness centrality term. As a demonstration, we present a Walk-a-Tron analysis of a discussion thread from ScienceBuzz based on the topic of composting. We scraped the content of the composting thread using a screen scraping script. For the most part, the posts in the composting discussion thread proceed sequentially from the initial post instigating the discussion and subsequent replies. However, the replies in this thread are made to individual posts and are not always presented in a chronological order. As such, a reply made to a post two years after the initial posting will appear adjacent to it on the page. Thus, the text mining work done accounts for the sequential display of posts across multiple ScienceBuzz pages—what a reader perusing the posts would see—as opposed to a strict through line from the beginning to the end of the discussion.

The top ten betweenness centrality terms from the composting thread[26] are presented in Table 6.2.

With these high betweenness centrality terms identified, we can now chart their topological distribution in the composting discussion thread. The results of a Walk-a-Tron chart can be viewed in Fig. 6.7.

Much of what Table 6.2 and Fig. 6.7 depict is to be expected about a discussion thread focused on composting. The terms "compost" and

Table 6.2 The top ten betweenness centrality terms from the composting thread

Token ID	*Betweenness centrality*
Compost	0.357854551409
Composting	0.0903304205664
Good	0.0899063098227
Think	0.0861110257402
Can	0.0802672637582
Help	0.076454309094
Dirt	0.0764441826518
Earth	0.0682752151787
Worm	0.0630377647016
Make	0.057143505189

"composting" prevail as do their cyclic walks. In this topographical visualization, the smaller the circumference circles, the more frequently the term repeats. Thus, the "compost" track that leads the visualization is characterized by numerous (relatively) tiny circles in rapid succession—especially in the early going. "Composting," which could be conflated with "compost" in terms of semantic function also occurs with frequency. However, what is interesting about "composting" as a term or node in the network is that it is followed closely (in betweenness centrality value) by "good," "think," and "can." Figure 6.8 depicts a zoomed-in view of a region of the Walk-a-Tron visualization in which "compost," "composting," "good," and "think" contain smaller-radii circles for each of the listed terms.

The region highlighted in Fig. 6.8 represents the range of tokens between 747–870. The natural language text of this region, reconstructed from the range of tokens, reads as follows:

> We hope to see more composting. I tiried this at home and it is so amazing my dad has a compost pile in the backyard and it helps fertilize the garden i think that sometimes compost smells but i guess it good for your garden. i have an garden in my backyaerd and i haveto touch dirty compost. I think that comnposting is very important! I'll try to do it at home. Earth for the win!!! :D worms make a type of compost to My mom cookies taste like compost Boooooooooooo Pollutioners Yes. That is the simplest solution to global warming and animal extinction. Don't pollute. worms are gross but good for composting   Composting saves the earth... DO IT!!!!! I think that compost is a really good thing to do, because it's kind of like you're returning what Mother Nature gave to you. compost is good for plants. ("Get all the Dirt on Composting")

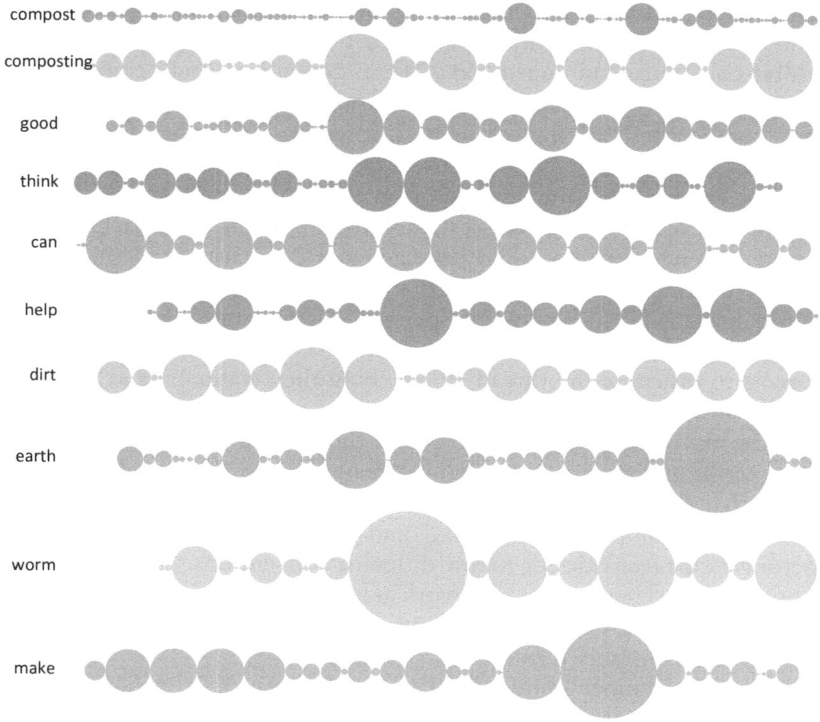

Fig. 6.7 Compost discussion thread walks

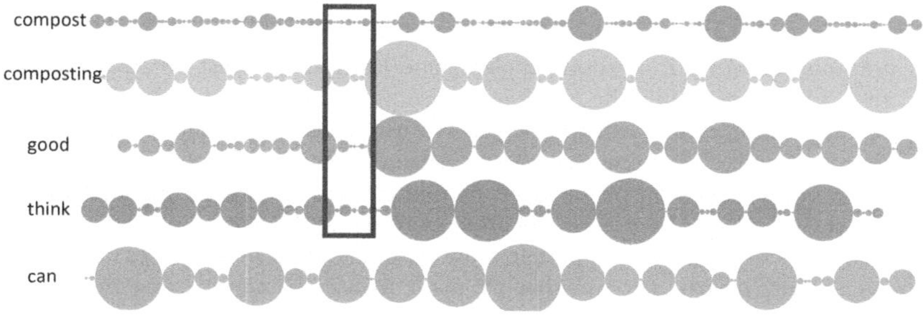

Fig. 6.8 Highlight of Walk-a-Tron visualization

This region of text, which contains multiple posts by different authors, exhibits some interesting and familiar attributes of discussion threads that exhibit coherence and engagement. First, “compost” and “composting” are uttered, keeping the topic of the thread in focus. Second, “compost” and “composting” are framed by both interpersonal hedging statements (“I think”) and endorsements (“I think that compost is a really good thing to do”). We note that the framing of “composting” in these ways continues throughout this snapshot, establishing an echolalic character to the thread. Taken together, we may be seeing genre signals sent by the participants: recurring discursive actions that bid for recognizability from other interlocutors while managing interpersonal relationships. By “managing,” we refer to a stance in which interlocutors make conspicuous their attentiveness to others in the thread by couching their statements as personal, not declamatory.

If the above analysis is plausible, and the overlap of betweenness centrality terms indicates genre signaling, then it should also be reasonable to expect that those regions that do not have high incidence and overlap of “compost,” “composting,” “think,” and “good” would portray different rhetorical characteristics or follow different conventions. In other words, the topological change in the network graph should imply a signal change. Figure 6.9 highlights a region notable for its lack of any of the top ten betweenness centrality terms.

This region comes quite a bit later in the thread. It includes tokens 1873–1989. The natural language text in this area of concern reads as follows:

> I think dirt is a very importent part of life, even if it can get a little messy, that’s my opinion. dirt is the source of EVERYTHING without soil our planet would be a empty husk of nothing i think that this world would be nothing without soil no food no plants and last but not least no crops and if we dont eat we could die A Service Learning Project must be incoporated into our community with the help of youth voices and adults contributing together and stopping the crisis. Working together as a community and decreasing the amount of waste left behind the soil can afffect our future if we act right now. There are many other factors that the soil we use can benefit the Earth. The affect of people liotering and leaving their messes behind can actually destroy the soil that we plant to breathe air or consume our bodies. Composing- Taking waste materials and mixing it with a component that benefits the earth.

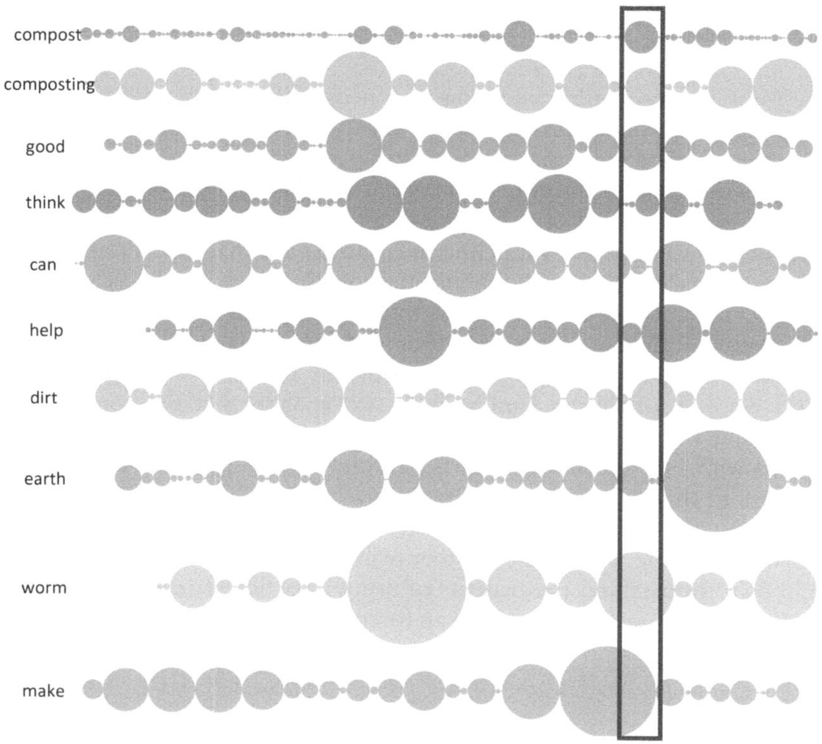

Fig. 6.9 Highlight of Walk-a-Tron lacking high betweenness centrality terms

In the section above, we see similar sentiments, but articulated in a different register. This section moves beyond "compost" or "composting" as the key term, focusing instead on "soil" and "dirt." Moreover, the positive valence signaled by "good" is replaced with urgent prognostications about the future sustainability of the earth. While this section of the composting discussion thread is still on topic, it differs significantly from those regions dwelling on the high betweenness centrality terms lexically and rhetorically. We would also argue that there has been a generic shift from hedgy affirmation of composting to the topics of pedagogy ("Service Learning Project") and "community building" mixed with some dire soothsaying. What is interesting is how clearly this shows up as an outlier post in the visualization, marking a clear shift in the discursive pattern typifying the rest of the thread.

Are You Being Topological Without Even Knowing It?

We want to close with something of a hedge of our own regarding the topological analysis of texts and discourse. While we hope that taking the time to articulate ideas that pertain to rhetoric and graph theory has produced some helpful insight, we expect that some might find all of this discussion a bit overblown. There are, after all, techniques such as latent dirichlet allocation topic modeling in common use that take a topological approach, performing multidimensional analysis of texts to find clusters of similar terms. These can be used without knowing too much about how the sorting algorithms that drive these analytic tools work. But what we also want to acknowledge is that it may not be all that interesting to some users who are otherwise familiar with text analysis that the computer is not seeing a text, per se, but rather evaluating a graph or comparing many graphs with one another to return a result. So why go to all the trouble we have in this chapter to describe the *why* with the *how*? We think there is value to cultivating affinity between the theoretical principles that drive our work and the analytical tools that help us to carry it out. Doing so can help us to understand the benefits of using a particular technique as well as the limits. And it is the best way to create new analytical tools of our own design.

Notes

1. Freedman and Medway. *Genre in the new rhetoric.*
2. Schryer, "Records as genre."
3. Schryer, "Records as genre."
4. Spinuzzi, *Tracing genres through organizations: A sociocultural approach to information design.*
5. Teston et al., "Public voices in pharmaceutical deliberations: Negotiating 'clinical benefit' in the FDA's Avastin Hearing."
6. Miller, "Genre as Social Action."
7. Miller and Shepherd, "Questions for genre theory from the blogosphere."
8. In this chapter, we use the word "topological structure" and "graph" simultaneously because a graph in mathematical terms is a topological structure. It is a collection of points connected by edges. It need not be a regular grid with edges at right angles as we might associate with graph paper or a grid. Any mess of points and lines will do to make a proper graph.

9. Omizo and Hart-Davidson. "Finding Genre Signals"; Omizo and Hart-Davidson, "Hedge-o-Matic."
10. Graham et al., "Statistical genre analysis: Toward big data methodologies in technical communication."
11. We highly recommend the fascinating (and eminently readable for the nonexpert) biography *Euler's Gem: The Polyhedron Formula and the Birth of Topology* by D. Richeson. The claim that topology as the study of "malleable objects" is relatively unknown is Richeson's. We agree with it. We also agree that it is a topic that deserves much broader recognition and study beyond the group of mathematics scholars who count themselves as topologists.
12. An important part of topological analysis involves understanding holes or voids.
13. Bakhtin, "Problems of Speech Genres," 91.
14. Pare, "Rhetorical Genre Theory and Academic Literacy."
15. Pare, "Rhetorical Genre Theory and Academic Literacy," A85.
16. Miller, "Genre as Social Action."
17. Schryer, "Records as genre."
18. Biesecker, "Rethinking the Rhetorical Situation from within the Thematic of 'Différance.'"
19. See Lemke, "Typological and topological meaning in diagnostic discourse"; Diesner and Carley, "Revealing social structure from texts"; Popping. "Knowledge graphs and network text analysis"; Tamboyang and Carley, "Network text analysis in computer-intensive rapid ethnography retrieval: an example from political networks of Sudan."
20. Pare, "Rhetorical Genre Theory and Academic Literacy."
21. Strapparava and Valitutti, "WordNet Affect: an Affective Extension of WordNet"; Neviarouskaya, Prendinger, and Ishizuka, "Textual affect sensing for sociable and expressive online communication."
22. Brin and Page, "Reprint of: The anatomy of a large-scale hypertextual web search engine."
23. Erkan and Radey, "LexRank: Graph-based lexical centrality as salience in text summarization."; Corman, Kuhn, and McPhee, "Studying Complex Discursive Systems."
24. Miller, "The Aristotelian Topos: Hunting for Novelty."
25. Pigg, Hart-Davidson, Grabill, and Ellenbogen, "Why People Care About Chickens."
26. "Get All the Dirt on Composting." *ScienceBuzz,* accessed October 15, 2016, http://www.sciencebuzz.org/topics/get-all-dirt-compost.

Bibliography

Bakhtin, M. 1986. "The Problem of Speech Genres." In *Speech Genres & Other Late Essays* ed. C. Emerson & M. Holquist, trans. Vern W. McGee, 60–102. Austin: University of Texas Press.

Biesecker, Barbara A. 1989. "Rethinking the Rhetorical Situation from within the Thematic of 'Différance." *Philosophy & Rhetoric* 22: 110–130.

Brin, Sergey, and Lawrence Page. 2012. "Reprint of: The Anatomy of a Large-Scale Hypertextual Web Search Engine." *Computer Networks* 56, no. 18: 3825–3833.

"Chicken and Egg." *ScienceBuzz*. http://www.sciencebuzz.org/blog/chicken-and-egg/. Accessed 3 Aug 2006.

Corman, Steven R., Timothy Kuhn, Robert D. McPhee, and Kevin J. Dooley. 2002. "Studying Complex Discursive Systems." *Human Communication Research* 28, no. 2: 157–206. doi:10.1080/0363775032000167406.

Diesner, Jana, and Kathleen M. Carley. 2005. "Revealing Social Structure from Texts." In *Causal Mapping for Research in Information Technology*, ed. V.K. Narayanan, 81–108. Hershey: Idea Group Publishing.

"Emotional Marketing Headline Analyzer." *Advanced Marketing Institute*. http://www.aminstitute.com/headline/. Accessed 25 October 2016.

Erkan, Günes, and Dragomir R. Radev. 2004. "LexRank: Graph-Based Lexical Centrality as Salience in Text Summarization." *Journal of Artificial Intelligence Research* 22: 457–479.

Freedman, Aviva, and Peter Medway, ed. 2003. *Genre in the New Rhetoric*. Bristol: Taylor & Francis.

"Get All the Dirt on Composting." *ScienceBuzz*. http://www.sciencebuzz.org/topics/get-all-dirt-compost. Accessed 15 October 2016.

Graham, S. Scott, Sang-Yeon Kim, Danielle M. DeVasto, and William Keith. 2015. "Statistical Genre Analysis: Toward Big Data Methodologies in Technical Communication." *Technical Communication Quarterly* 24, no. 1: 70–104.

Lemke, Jay L. 1999. "Typological and Topological Meaning in Diagnostic Discourse." *Discourse Processes* 27, no. 2: 173–185.

Miller, C.R. 1984. "Genre as Social Action." *Quarterly Journal of Speech* 70, no. 2: 151–167.

———. 2000. "The Aristotelian Topos: Hunting for Novelty." In *Rereading Aristotle's Rhetoric*, ed. Alan G. Gross and Arthur E. Walzer, 130–146. Carbondale/Edwardsville: Southern Illinois University Press.

Miller, C.R., and D. Shepherd. 2009. "Questions for Genre Theory from the Blogosphere." In *Genres in the Internet: Issues in the Theory of Genre*, ed. Janet Giltrow and Dieter Stein, Vol. 88, 263–290. Amsterdam: John Benjamins Publishing.

Neviarouskaya, Alena, Helmut Prendinger, and Mitsuru Ishizuka. 2007. "Textual Affect Sensing for Sociable and Expressive Online Communication."

In *International Conference on Affective Computing and Intelligent Interaction*, 218–229. Berlin/Heidelberg: Springer.
Omizo, Ryan, and William Hart-Davidson. 2016. "Finding Genre Signals in Academic Writing." *Journal of Writing Research* 7, no. 3: 485–509.
Omizo, Ryan and William Hart-Davidson. "Hedge-o-Matic." *Enculturation* 7. http://enculturation.net/hedgeomatic. Accessed 1 August 2016.
Paré, Anthony. 2014. "Rhetorical Genre Theory and Academic Literacy." *Journal of Academic Language and Learning* 8, no. 1: A83–A94.
Pigg, Stacey, William Hart-Davidson, Jeffrey Grabill, and Kirsten Ellenbogen. 2016. "Why People Care About Chickens and Other Lessons About Rhetoric, Public Science, and Informal Learning Environments." In *Science & the Internet: Communicating Knowledge in a Digital Age*, ed. Alan G. Gross and Jonathan Buehl, 247–266. Mahwah: Baywood.
Popping, Roel. 2003. "Knowledge Graphs and Network Text Analysis." *Social Science Information* 42, no. 1: 91–106. doi:10.1177/0539018403042001798.
Richeson, David S. 2012. *Euler's Gem: The Polyhedron Formula and the Birth of Topology*. Princeton: Princeton University Press.
Schryer, Catherine F. 1993. "Records as Genre." *Written Communication* 10, no. 2: 200–234.
Spinuzzi, Clay. 2003. *Tracing Genres Through Organizations: A Sociocultural Approach to Information Design*. Vol. 1. Cambridge, MA: MIT Press.
Strapparava, Carlo, and Alessandro Valitutti. "WordNet Affect: An Affective Extension of WordNet." In *LREC*, vol. 4, 1083–1086. Lisbon.
Tambayong, Laurent, and Kathleen M. Carley. 2012. "Network Text Analysis in Computer-Intensive Rapid Ethnography Retrieval: An Example from Political Networks of Sudan." *Journal of Social Structure* 13, no. 2: 1–24.
Teston, Christa B., S. Scott Graham, Raquel Baldwinson, Andria Li, and Jessamyn Swift. 2014. "Public Voices in Pharmaceutical Deliberations: Negotiating "Clinical Benefit" in the FDA's Avastin Hearing." *Journal of Medical Humanities* 35, no. 2: 149–170.

CHAPTER 7

Mapping Rhetorical Topologies in Cognitive Neuroscience

Jordynn Jack, L. Gregory Appelbaum, Elizabeth Beam, James Moody, and Scott A. Huettel

Hardly a week goes by in which a news item does not appear about how neuroscientists are getting closer to "mapping" the human brain. Indeed, the idea that the brain is *terra incognita*, awaiting scientific explorers, is itself a common *topos* used to describe what neuroscientists do. The comparison is often explicit; as Shelly Fan puts it:

J. Jack (✉)
Department of English and Comparative Literature, University of North Carolina, Chapel Hill, NC, USA

L.G. Appelbaum
Department of Psychiatry and Behavioral Sciences, Duke University School of Medicine, Durham, NC, USA

E. Beam
Stanford University School of Medicine, Stanford, CA, USA

J. Moody
Department of Sociology, Duke University, Durham, NC, USA

S.A. Huettel
Department of Psychology and Neuroscience, Duke University, Durham, NC, USA

© The Author(s) 2017

L. Walsh, C. Boyle (eds.), *Topologies as Techniques for a Post-Critical Rhetoric*, DOI 10.1007/978-3-319-51268-6_7

> Back in the Age of Discovery, rudimentary maps allowed European explorers to sail into the vast unknown. They began charting new worlds, and in turn, made newer maps that helped future generations better understand the lands and seas that cover our world. Now, thanks to a new—if slightly different—type of map, we may be approaching a new age of discovery.[1]

Yet, this task is much more complex than the phrase "mapping the brain" suggests. Neuroscientists are not simply interested in identifying structures of the brain—its mountains and valleys—but also in describing the connections between brain structures and the psychological functions that can be attributed to those structures and connections. While attempts to do so date back centuries, often using information gained from brain surgeries or lesions to attribute brain functions to regions, today the state of the art involves imaging techniques, especially functional magnetic resonance imaging (fMRI). To date, multiple maps of the brain have been generated, from Korbinian Brodmann's histological 1909 map, which identified 52 areas, to the newest map, generated using data from the Human Connectome Project, which identified 180 areas per brain hemisphere.[2] These attempts to map the brain might be understood as descriptions of its topography. But what of its *topology*?

Because the discipline of cognitive neuroscience draws connections between biological systems and psychological concepts, it offers an interesting case study for topological rhetorics. This article describes how our research team has attempted to create a map of the rhetoric of the brain, one that charts not the brain regions and coordinating psychological concepts as they are, but as they are established rhetorically in research articles. That is, our maps delineate the relationships scientists seek to establish when publishing experimental data—relationships that take on greater or lesser persuasive strength as they are made again and again in published studies.

As an interdisciplinary team of researchers, we developed an approach to understanding this topology that is sensitive to the rhetoricity of cognitive neuroscience publications in which data appear. Our model provides a topology of *how neuroscience represents the relationships between brain regions and concepts in scientific publications.* In this essay, we describe how we curated a corpus of abstracts drawn from neuroscience articles published in five top cognitive neuroscience journals over a span of two and a half years. Then, using network analysis, we generated a topology of brain regions and concepts as they are described in the collection of abstracts. This topology might be understood, in rhetorical terms, as demonstrating the special topics that make up the discipline of cognitive neuroscience—the set of discipline-specific terms through which researchers generate hypotheses and interpret results. Next, using computational discourse analysis tools, we suggest what common *topoi* are used to discursively establish those connections.

Finally, we describe how our semantic model also functions as a kairology—a map of opportunity for neuroscience researchers. This type of map can help researchers to identify underresearched areas (such as the thalamus) that represent kairotic spaces for intervention or, conversely, areas that are oversaturated (such as the visual cortex), making it difficult to add new or exciting data. It also helps researchers to identify "islands"—cases where subdisciplinary focal areas, such as "neuroeconomics," have not yet been integrated with the broader field, but where doing the rhetorical work to do so—drawing connections in literature reviews or within research articles between this area and the broader network—could yield great benefit.

To conclude, we suggest how our approach might be applied to other cases—how using social network analysis (SNA) as a tool for rhetorical analysis does not simply mean describing topics that appear together (which is how SNA is often used in rhetoric and discourse analysis), but as a tool for understanding how disciplinary topologies form over time and for identifying kairotic opportunities within a discursive field.

Background

The study and cataloging of *topoi* has a long history in rhetoric, forming the backbone of invention techniques described in Aristotle's *Rhetoric* and onward. As Carolyn A. Miller has argued, *topoi* can serve problem-solving functions for rhetors: "Viewing a problem from the vantage of a *topos*, so to speak, can reveal or make possible new combinations, patterns, relationships that could not be seen before."[3] Studying the *topoi* used in a particular field of discourse allows rhetoricians to understand how rhetors draw on *topoi* to generate new arguments. Historically, *topoi* have been understood to include two types: *koinoi topoi*, or general topics, and *eide*, or special topics. As Richard C. Huseman puts it, the *konoi topoi* include "topics which suggest form for arguments" while the *eide* include "those which suggest material and ideas for arguments."[4] While scientists use common topics in their reasoning, they also employ special topics, which are particular to a field of discourse or discipline. As Miller describes it, special topics have three potential sources: "conventional expectation in rhetorical situations, knowledge and issues available in the institutions and organizations in which those situations occur, and concepts available in specific networks of knowledge (or disciplines)."[5] Thus, special topics are closely linked to disciplinary knowledge; that is, training in a scientific discipline embeds rhetorical training in the form and content of argument within that field.

Within scientific disciplines, for instance, Lawrence J. Prelli outlines special topics (or *eide*) such as accuracy, internal consistency, external consistency, scope, simplicity, elegance, and fruitfulness.[6] These, and other special topics, help researchers across a variety of disciplines to develop persuasive arguments for their findings. While some of *topoi* are present in most (if not all) scientific disciplines, Walsh argues that outlining the *topoi* specific to STEM fields can "illuminate the communal organisms that are STEM fields" as well as the "topical signatures" of specific disciplines.[7]

Here, we outline some unique "topical signatures" of cognitive neuroscience, a signature forged in part by the interdiscipinarity of the field and the specific apparatus (including instruments such as fMRI) it uses to connect psychological concepts with anatomical regions of the brain. We describe some of the special topics as well as variations on the common topics that have been developed in the field of cognitive neuroscience.

More broadly, though, our article demonstrates one methodology by which we can construct such a topical signature that shows not only which *topoi* are present but also how they are related. While scholars have largely approached rhetorical topology by listing *topoi*—beginning with the 28 *topoi* outlined in Aristotle's *Rhetoric* and expanding them to account for particular fields of discourse—they have done less work to illuminate how *topoi* connect with each other. Nonetheless, contemporary understandings of *topoi* clearly suggest that they might be understood as networks. Indeed, Miller defines a *topos* "a point in semantic space that is particularly rich in connectivity to other significant or highly connected points."[8] Using social network analysis and a large corpus of texts, it is possible to map those connections and gauge the connectivity of particular *topoi*.

Methods

As Walsh and Andrew B. Ross describe it, "A topical approach reminds the critic to enter inductively into a discourse, to watch for the commonplaces that stakeholders resort to as they form arguments, and to employ these discovered commonplaces as a foundation for theorizing the power dynamics of that discourse."[9] One way to do this, of course, is to employ close reading and coding of scientific texts, an approach both Walsh and Ross employ in their work. Walsh's call for topical analysis involves "certain sampling procedures via its insistence on close reading of multisentence and even multiparagraph spans of text."[10] While this approach has the advantage of granularity, it does not allow researchers to examine a very large number of texts. Accordingly, we used social network analysis and computational discourse analysis (CDA) software to analyze a wider range of texts—in this case, abstracts from research articles. We were confident that we could capture most instances of the special *topoi* in question because there is a limited, specific set of terms used to describe anatomical regions and psychological concepts. Our computational tools helped us to generate a list of these terms and to determine their frequency and connectedness to other terms. Moreover, this approach has the advantage of helping to show the connections between *topoi* as well as the power dynamics of the field, indexed by our analysis of which topics become most highly connected to others and which become isolated.

To identify the common and special topics used within cognitive neuroscience, we gathered every article within a span of two and a half years (January 1, 2008, to June 30, 2010) from five leading journals: *Nature Neuroscience, Neuron, Journal of Cognitive Neuroscience, NeuroImage,* and *Journal of Neuroscience.* From the resulting 7675 abstracts, we selected those that met the following criteria:

1. They used functional magnetic resonance imaging (fMRI) as the primary form of data collection. We chose fMRI due to its popularity, as the most commonly used technique in our corpus, and eliminated other techniques in order to ensure that the abstracts we selected would have similar terminology.
2. The primary goal of the research was to understand links between the human brain and some psychological function. We applied this criterion in order to ensure that the research we included could be clearly considered as cognitive neuroscience. We excluded animal

studies because human and animal brains differ in anatomical organization and behavior.
3. The article reported empirical data. We applied this criterion in order to minimize the confounding effects of metanalyses and review articles, which perform a different rhetorical and scientific function from primary research articles.

After discarding articles that did not meet these criteria, we were left with a corpus of 1127 studies. The corpus included both the title and the abstract for each article. Next, we generated word frequency lists from the corpus and selected the 100 most common anatomy terms (such as *hippocampus* or *fusiform face area*) and concept terms (such as *memory* or *risk*). We generated a thesaurus to consolidate similar terms, plurals, and acronyms. In addition, we combined frequently occurring bigrams, which increased the granularity of the terms. For example, *working memory* is a bigram that is narrower than *memory*; *left inferior frontal gyrus* is a combination of at least 2 bigrams, resulting in a more specific anatomical region than *frontal gyrus.*

Then, we used Automap, a social network analysis software, to create three different networks: conceptual (reflecting connections between concept terms, such as *memory* to *representation*), anatomical (representing connections between brain structures, such as *prefrontal cortex* to *hippocampus*), and functional (concept to anatomy and anatomy to concept). We used Organization Risk Analyzer software to visualize those networks.[11]

Special Topics

The three maps we generated demonstrate how cognitive neuroscience organizes itself around special topics regarding cognitive concepts, brain structures, and the interrelationships between them. To reiterate, our maps do not describe the brain's physical terrain, but the rhetorical terrain researchers have plotted as they work, through experiment after experiment, to solidify connections between brain regions and psychological concepts as accepted knowledge within the discipline. Because our networks map *discourse* (namely in research article abstracts), what we see here is not a depiction of how psychological concepts might actually be organized in the brain, but of how researchers rhetorically fit them together in constructing experiments and interpreting results, and how they situate their research with relation to other research in the field.

Conceptual Topoi

The special topic of concepts draws on the discipline of psychology and the set of terms it has developed to describe human cognition. Our conceptual network (Fig. 7.1) suggests three main categories under which sub-*topoi* might be grouped: perception/attention, representation/memory, and cognition/control. As we have argued in our previous report, these categories "recapitulate[] the long-standing division of the mind into stages of information processing: perceiving something, representing

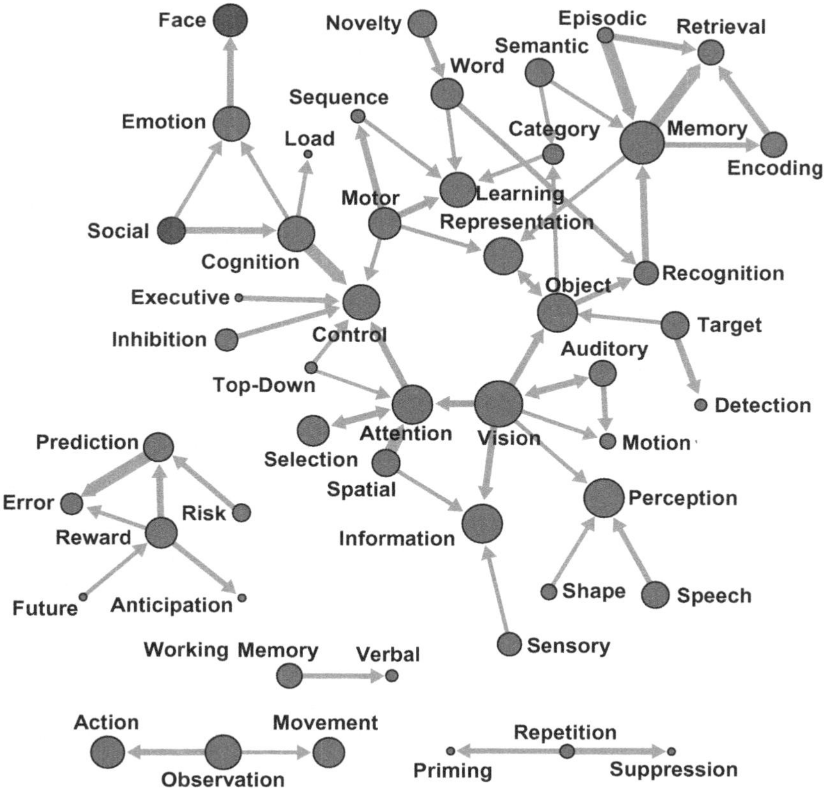

Fig. 7.1 Network of conceptual *topoi* in 1127 neuroscience research article titles and abstracts

it in memory, and then controlling behavior accordingly."[12] In rhetorical terms, we might say that these three stages may function heuristically for researchers, helping them to focus experimental designs and interpret the results of those experiments as they pertain to one of those special topics.

The arrows between these *topoi* show how researchers rhetorically link those concepts together. Some links are well trod (shown with a thicker arrow), such as the link between vision and information or cognition and control. In other words, those *topoi* to appear in frequent proximity to one another in research abstracts. From the map, it is also possible to ascertain clusters of *topoi* that are likely to appear together in an abstract; these clusters mark concepts that researchers understand as highly related.

From a rhetorical perspective, we might assume that researchers may work from one *topos* to another that is frequently connected to it. For instance, a researcher interested in the *topos* of "memory" might be likely to describe their study and its significance by employing related *topoi* of "retrieval" or "encoding." From this area of the map, we can also learn which *topoi* are related, but not connected. For instance, we see that researchers are unlikely to discuss both episodic and semantic memory in the same abstract, since those two terms are not themselves connected in the network. Instead, we learn that researchers are likely to focus their experiment and explanation of results on one or the other.

The network also helps rhetoric researchers to distinguish between the frequency of a *topos* and the strength of its relationship to other *topoi*. In our maps, the frequency of a *topos* is indicated by the diameter of the circle used to represent it. A *topos* can be relatively frequent without necessarily being linked strongly to other *topoi* in the network. For instance, in our map, the *topos* "information" is relatively frequent, but not especially densely connected to other *topoi*.

Finally, this topology suggests that some *topoi* are relatively isolated from the main network, such as the action/observation/movement triad or the set of concepts related to risk and reward. As rhetoricians, we might ask why this is the case. The risk/reward cluster, for instance, describes research from the emerging subdiscipline of neuroeconomics; our map suggests that this area of inquiry was (at least at the time represented in our corpus) relatively isolated from the larger field. Perhaps, researchers in that area of focus were concentrating more on establishing a set of special *topoi* that could be used to form that specialized area of inquiry. We might hypothesize that, over time, those concepts might become more deeply connected with others in the network, assuming researchers in neuroeconomics are able to establish their findings as central to the field as a whole. Or, to put it in rhetorical terms, special topics common in neuroeconomics had not yet infiltrated the broader field of cognitive neuroscience during the period we studied.

Anatomical Topoi

Whereas our conceptual map revealed three subdivisions, the anatomical map (Fig. 7.2) was instead dominated by three specific nodes: *prefrontal cortex (PFC), amygdala,* and *anterior cingulate cortex (ACC)*. The frequency of these terms reflects the significance researchers attach to them. As we will see in the functional map, the PFC is commonly implicated in planning, decision-making, and rational thought; the *amygdala* is implicated in emotion processing, and the ACC in decision-making and control. As special topics, brain regions are difficult to separate rhetorically from the concepts that have become attached to them. Certain regions become "hot" or "sexy" for researchers because of these connected *topoi*. Researchers may be likely to target these areas as regions of interest (ROIs) in their studies because they carry such rhetorical significance in the field; or, possibly, results pertaining to those regions may be more persuasive and interesting to journal editors and peer reviewers.

It might simply be the case that these regions appear more frequently because they are most important for human cognition. However, our analysis also revealed that two terms were disproportionately central in the map compared to their frequency in the corpus: *thalamus* and *insula*. These latter regions seem to occur in the literature with relation to a wide range of other regions, yet they do not tend to receive as much attention from researchers as the prefrontal cortex, amygdala, and ACC. Thus, it is possible that researchers, when developing hypotheses, designing experiments, and then reporting results, are drawn to brain regions that seem to have exceptional status as special topics in the field—places that have developed over time as reliable ones from which to generate compelling arguments with relation to concepts seen as key to understanding human cognition. Other regions, such as the *thalamus*, may have less weight as special topics precisely because they can be linked to many different concepts, or because they are linked to concepts that seem less compelling to researchers.

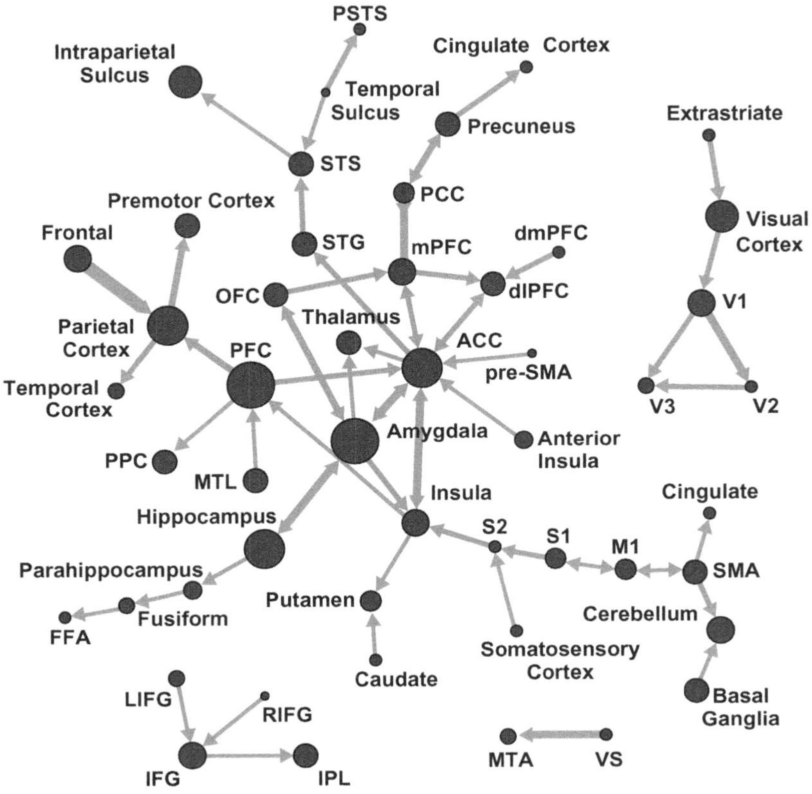

Fig. 7.2 Network of anatomical *topoi* in 1127 neuroscience research article titles and abstracts

Functional Topoi

Our third network (Fig. 7.3), illustrating connections between anatomical regions and psychological concepts, is perhaps most interesting for a topological analysis because it suggests how two different types of special topics become linked. As was the case with the anatomical map, several key regions dominated the network: the prefrontal cortex, amygdala, the parietal cortex, and the hippocampus. Each of these regions forms a hub around which several conceptual terms are arrayed. For instance, the hippocampus is linked to novelty, memory, and learning in our corpus.

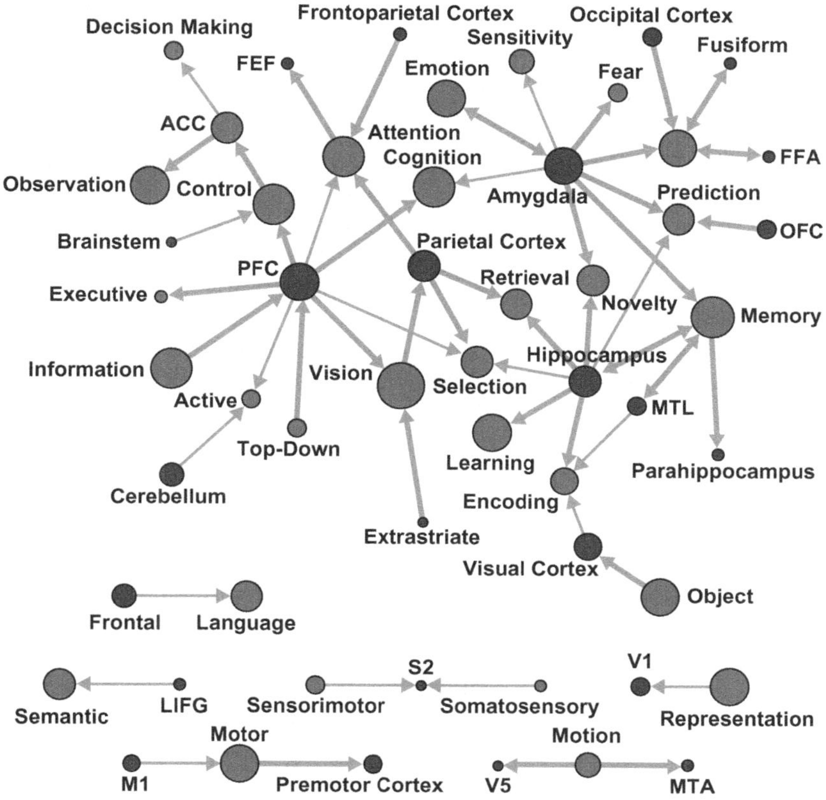

Fig. 7.3 Network of functional *topoi* in 1127 neuroscience research article titles and abstracts

From a rhetorical perspective, this suggests that certain anatomical regions become linked to certain psychological concepts through a process of accretion. Some of those links may themselves become special *topoi*, or generally accepted premises for arguments. For instance, the amygdala is generally understood for its association with emotion. However, only certain kinds of arguments are considered permissible within the field of cognitive neuroscience. For instance, one would think that activation in the amygdala might allow researchers to make claims about emotion, fear, novelty, or attention; activation in the PFC might allow for arguments about control, information, or vision. This type of reasoning has been dubbed "reverse inference" by the prominent neuroscientist Russell Poldrack; these arguments are those in which one attempts to "infer the engagement of particular cognitive functions based on activation in particular brain regions."[13] The validity of these arguments is questionable because any one brain region might be associated with a range of psychological concepts. Instead, Poldrack argues that experiments support reasoning in the opposite direction; these arguments take the following form: "if cognitive process X is engaged, then brain area Z is active."[14] Our map does not make it possible to determine whether the research abstracts we studied were using the reverse inference fallacy, since it only indicates proximity of two terms in a sentence.

However, Poldrack and others have argued that metanalyses can allow for some arguments to be made using reverse inference. Poldrack argues that "reverse inferences can suggest novel hypotheses that can then be tested in subsequent experiments," especially when based on maps that aggregate data from multiple studies, such as the BrainMap database (http://www.brainmap.org).[15] Our maps are different because they are based not on research data per se, but on how those data are described in research article abstracts. Nonetheless, our maps might suggest possible lines of inquiry for researchers who may notice the repeated co-occurrence (or lack thereof) of a given brain region and concept in the literature and design a new experiment to further strengthen evidence.

Despite the fallacy of reverse inference, our maps also demonstrate the persistence of brain structures that have been labeled based on their functional properties. For example, the "fusiform face area" is the part of extrastriate cortex that is more active when viewing faces.[16] At first, the practice of naming brain regions according to functions was relatively limited to vision neuroscience, where associations between anatomy and cognitive processes could be reliably evoked in single subjects, and the reverse inference was more likely to be accurate. However, even the fusiform face area has been found to be involved in processing a range of other stimuli, and the field has generally moved toward models of functional networks as opposed to functionally defined regions. It appears that the next wave of functional labeling has occurred in the naming of resting state connectivity networks. For example, the "frontoparietal control network" includes regions that earlier fMRI activation studies found to be involved in executive control.[17] With reverse inferences built into the language of cognitive neuroscience, it appears that despite increasing recognition of the reverse inference fallacy, researchers continue to reinforce that fallacy implicitly in the naming of regions and networks.

Overall, our social network analysis of a corpus of research article abstracts provides a snapshot of how researchers have made those links during a given stretch of time. Or, in rhetorical terms, we might say that our maps provide a meta-level view of the special topics used in the discipline and how those topics are connected.

Common Topics

The social network analysis we have conducted thus far does not extend to the common topics researchers use to present and interpret their data. Future extensions of this project may include common topics by including additional terms—specifically meta-discourse terms—as a variable. Linguists Ken Hyland and Polly Tse define meta-discourse as "the linguistic resources used to organize a discourse or the writer's stance towards either its content or the reader."[18] Within scientific discourse, meta-discourse commonly includes hedgers, used to downplay the impact of a statement; boosters, used to intensify the impact of a statement;[19] and reporting verbs and nouns that are used to describe the results of experiments.[20]

The most common hedges and boosters in our corpus do not differ dramatically from those provided in other studies of STEM discourse, although certain expressions identified by Hyland do not appear in our corpus or appear only rarely. Overall, the more moderate terms tend to be prevalent in our corpus among both hedges and boosters. Among the boosters, the most decisive terms (such as "no doubt" or "obviously") do not appear at all (Table 7.1).

More dramatic differences occur in our list of reporting verbs, many of which do not appear in the list provided by Joel Bloch (shown in bold in Table 7.2; words in italics do not appear in Bloch's list). This suggests that cognitive neuroscience may have developed a unique set of reporting verbs as part of its "topical signature." Moreover, we also identified a parallel series of nouns used to indicate results (shown in Table 7.3), which we are calling "reporting nouns."

Reporting nouns are used to describe what it is that researchers are finding when they use fMRI—"activations" or "processes" or "connectivity" are all variations of this term.

Because the field of cognitive neuroscience is predicated on establishing links between two different disciplines, we hypothesize, it privileges a set of *topoi* used to establish connections between anatomy and

Table 7.1 Hedges and boosters in a cognitive neuroscience corpus

Hedges		*Boosters*	
suggest	1249	show	1795
may	987	significan*	974
indicate*	481	find	929
about	436	specific	775
could	239	demonstrate	579
might	199	significant	507
would	189	importan*	463
rather	185	significantly	426
appear*	176	known	363
expected	156	important	314
possible	139	novel	300
unclear	138	new	268
often	136	specifically	246
likely	113	direct	200
typically	102	even	193
relatively	97	stronger	173
should	88	strongly	168
mainly	77	critical	157
largely	75	strong	148
estimate*	73	think	147
generally	72	directly	121
seem/s	69	functionally	113
assume	63	particularly	110
possibly	57	positively	109
argue*	56	establish	102
apparent	50	clear	65
around	49	importantly	63
typical	36	evident	50
usually	36	must	50
approximately	34	certain	42
in general	34	vrai	42
frequently	33	clearly	40
uncertain	27	believe	30
presumably	26	know	30
tend to	23	considerable	30
perhaps	22	always	26
almost	21	indeed	25
couldn't or could not	21	actually	19
probably	21	never	15
postulate*	20	prove	14
mostly	19	considerably	9

(*continued*)

Table 7.1 (continued)

Hedges		*Boosters*	
claim*	18	in fact	7
sometimes	18	obvious	5
plausible	13	really	3
felt	12	truly	3
apparently	10	definite	2
unlikely	10	beyond doubt	0
broadly	9	certainly	0
probable	9	conclusively	0
to our knowledge	8	decidedly	0
essentially	7	definitely	0
doubt/s	6	doubtless	0
feels	6	evidently	0
fairly	5	incontestable	0
guess	5	incontrovertible	0
quite	5	incontrovertibly	0
suppose*	5	indisputable	0
roughly	4	indisputably	0
somewhat	4	no doubt	0
wouldn't/would not	2	obviously	0
certain amount	1	of course	0
in most cases	1	realize	0
in our view	1	sure	0
maybe	1	surely	0
certain extent	0	undeniable	0
certain level	0	undeniably	0
doubtful	0	undisputably	0
from my perspective	0	undoubtedly	0
from our perspective	0	without doubt	0
from this perspective	0		
in most instances	0		
in my opinion	0		
in my view	0		
in this view	0		
on the whole	0		
ought	0		
plausibly	0		
suspect	0		
uncertainly	0		
unclearly	0		

* is used to show that we searched for that root and included all variations in our search (for example, estimate* includes estimates, estimated, etc.)

Table 7.2 Results-oriented verbs

Reporting		*Connecting*		*Activating*	
show	1933	*associate*	1248	*involve*	1279
suggest	1615	*correlate*	1075	*decrease*	1251
increase	1251	*reflect*	425	*activate*	870
find	930			*modulate*	412
observe	877			*engage*	378
compare	846			increase	377
predict	840			*induce*	298
investigate	793			*represent*	252
reveal	731			*elicit*	246
support	685			*is/was sensitive to*	241
demonstrate	582			*recruit*	234
measure	566			*implicate*	217
examine	543			*mediate*	213
identify	537			*evoke*	205
provide	527			*produce*	202
indicate	484			*exhibit*	193
state	456				
reflect	425				
report	367				
known	363				
determine	208				
propose	185				
contribute	183				
think	147				
hypothesize	141				
conclude	115				
describe	105				
establish	102				
consider	88				
discuss	88				
explain	88				
resulted	67				
assume	63				
argue	56				
is/was seen in	46				
imply	36				
believe	30				
postulate	20				
note	19				
claim	18				
prove	14				
mention	6				
point out	0				

Table 7.3 Reporting nouns

activation	3498
process/es/ing	3018
result/s	1497
network/s	1312
effect/s	1228
signal/s	968
finding/s	920
change/s	890
level	877
differences	836
representation/s	753
pattern	744
model/s	689
mechanism/s	620
analysis	576
connectivity	549
evidence	544
structure	432
interaction	429
paradigm/s	363
influence	313
correlations	303
circuit/sry	300
neural correlates	202
involvement	193
relationship	187
substrate/s	172
recruitment	132
strength	106
component of	54
associates	13

psychological concepts. These include verbs and nouns that are used to indicate connections and activations, and nouns that evoke neurological systems. These draw from the common topics but our analysis (using CDA) shows a specific lexicon used in the field to make arguments connecting anatomy to function: connecting verbs and nouns, activating verbs, and nouns that evoke neurological systems.

Connecting verbs include *reflect*, *associate*, and *correlate;* connecting nouns include *correlations*, *neural correlates*, and *relationship*. As common topics, these terms relate most closely to Aristotle's third common topic

(KT3), correlatives.[21] In cognitive neuroscience, these are used to posit connections between observed activity in brain regions and experimentally induced concepts without overstating any line of causality. Activating verbs include *activate, involve, modulate, induce, recruit, implicate, mediate,* and *evoke*. Activating nouns include *activation, processing/process,* and *signal*. This *topos* corresponds to Aristotle's 24th common topic (KT24), cause and effect, and is used to posit a stronger causal argument. Finally, nouns that evoke neurological systems include *network, model, mechanism, structure,* and *circuitry*. This *topos* is used to posit an underlying neurological system that accounts for the activations demonstrated in an experiment, and corresponds to Aristotle's tenth common topic (KT10), induction.

According to Walsh, the common topics can be divided into "comparative, concerned with difference and identity; dimensional, concerned with continuities and relative positions along continua; and causal, concerned with antecedence and consequence."[22] We might group hedgers and boosters as dimensional common topics and reporting verbs and nouns as causal common topics; used in combination, these meta-discourse terms are used to either strengthen or weaken the writer's confidence in a causal relationship between an experimental situation that evokes a specific concept (say, memory or vision) and activation in a specific brain region. For example, in this sentence a combination of hedges and boosters are used to make a causal relationship seem stronger: "the tendency of an individual to wait for a larger, delayed reward **correlates directly with** BOLD signal in the lateral orbitofrontal cortex."[23] In contrast, here is an instance of hedges and boosters used to make a causal relationship seem weaker: "The right inferior orbitofrontal cortex, pre-supplementary motor area, anterior cingulate, and posterior cingulate showed similar patterns of activation with both regret and rejoice, suggesting that these **regions may be associated with** surprise from the realization of relatively unlikely events."[24] We might consider these terms to be causal only in a loose sense, however, since cognitive neuroscientists only posit associations between experimentally induced concepts and brain regions. Indeed, the term "cause" appears only 31 times in the corpus.

While a causal relationship is carefully avoided, researchers do assume that underlying structures support the correlations they have identified. Boettiger *et al.*, for instance, state: "These results **shed new light on** the **neurobiological underpinnings** of temporal discounting behavior;"[25] Chandrasekhar *et al.*, state that "Our results suggest that distinct, but overlapping **networks** are **involved in** the experiences of regret and rejoice."[26]

In lieu of causal relationships, then, researchers posit neurological structures; this move seems closest to the common topic of induction (KT10).

While Walsh identifies these three common topics (correlatives, cause and effect, and induction) as most commonly appearing in the results and discussion sections of research articles, we see them here, in research article abstracts, as well. Since abstracts tend to capitulate the structure of research articles, this is not surprising. Indeed, these *topoi* tend to appear in the last few sentences of an abstract, where writers are summarizing the results of a study and its significance.

Conclusion

Our research shows that computational tools, including social network analysis and computational discourse analysis, can provide a useful overview of a field of scientific discourse and should be included among the research methods used to study rhetorical topologies. Such research benefits from an interdisciplinary approach; in our case, we drew upon expertise from researchers in cognitive neuroscience, sociology, rhetoric and composition, and English.

The topology we generated functions not simply as a map of how the field is, but as a map of opportunity for researchers. For one, researchers might attend to the "islands," or small groups of terms that are disconnected from the main body of research. The presence of these islands may suggest that a new subfield, such as neuroeconomics, has been flourishing on its own and that the time may be ripe for researchers to connect that work to the main network. Another opportunity lies in key terms that are highly central, if less frequent, in the corpus. In the anatomical network, these included *insula* and *thalamus*; in the conceptual network they included *selection, emotion,* and *control*. These terms mark examples of topics that are underrepresented in the research.

While our maps are useful for researchers in cognitive neuroscience, they also represent opportunities for scholars in the rhetoric of science. A rhetorician might be interested in analyzing how a given brain region or psychological concept became dominant in the field—or how a once-dominant *topos* may have lost its cachet. This type of research would involve more traditional methods, such as close reading, to identify the specific contexts in which such gains and losses have occurred, as well as computational methods to chart the rise and fall of a *topos* over time. One might also be interested in connecting the texts used here (research article abstracts) with a broader range of texts. For example, to study the field of neuroeconomics, it would be interesting to note relationships between scientific texts, popular books, news sources, and websites, all of which help to circulate *topoi* that model decision-making using economic metaphors. Another interesting approach would be to study how neuroscience *topoi* circulate within other academic disciplines, a project that has begun to be undertaken by researchers interested in neurorhetorics as well as researchers in neuroscience.[27] For example, David R. Gruber has studied how the *topos* of "mirror neurons"—one that is contested within the field of neuroscience—has been taken up (typically without debate) in humanities and social science fields.[28] While typically undertaken using methods of close reading and rhetorical analysis, such projects could also draw upon computational methods like the ones we have used here in order to provide a more comprehensive view of how *topoi* move between disciplines. Along these lines, Clement Levallois, John A. Clithero, Paul Wouters, Ale Smidts, and Scott A. Huettel have used digital mapping techniques to study the subdiscipline of neuroeconomics, examining how conceptual terms (or special *topoi*) from different disciplines (including neuroscience, economics, and biology) interanimate in research articles. Rhetoric scholars might be interested in such approaches because they can help to map the "topical signatures" of interdisciplinary communities, not just single disciplines, showing how *topoi* become linked and unlinked as they move across communities.[29]

As with any topology, computational approaches are limited by methodological decisions and constraints. The semantic classes in our analysis were selected *a priori* by the researchers, as were the terms belonging to each class, by contrast to approaches such as topic modeling which identify key terms in an unbiased manner on the basis of their semantic relations.[30] Furthermore, in our case, the corpus we used was drawn solely from research abstracts, which are themselves persuasive summaries of experiments and should not be considered equal to the experimental results themselves. Due to their brevity and their persuasive function, abstracts may permit less subtlety than research articles themselves. It is possible that researchers highlight certain findings (certain brain regions or concepts) in abstracts that they consider more prominent or interesting to the field. Thus, a corpus drawn from the full text of research articles might provide different results.

In addition, there are particular kinds of interpretations and assumptions that are routinely made in abstracts. For one, abstracts frequently refer to the psychological processes that the authors intended to study, without typically naming the task that was used to probe those processes. As an example, "interference" appeared in the corpus 61 times, while "Stroop" (referring to the Stroop Test, commonly used as a task to model interference) appeared only 21 times. It may be more or less important to specify the task depending on how well it has been validated and how it is interpreted.[31]

Further, our analysis does not account for what might be termed the "not problem." That is, we have not eliminated from our results instances where researchers have found a lack of correlation between a brain region and concept, between two brain regions, or between two concepts. For instance, our analysis would not pick up on the disassociation in this excerpt from an abstract: "Furthermore, the right VLPFC but not the SPL showed the greatest activation during the nogo decision trials. This suggests both a functional dissociation between these areas and a role for the right VLPFC in rule-guided inhibition of behavior."[32] Here, the authors are interpreting their results to show that one brain region, but not another, was activated during the experimental decision. Our method, to date, is not sensitive to this distinction. Similarly, our approach glosses over any relational qualifiers, like "weak" or "strong." Thus, a computational approach to rhetorical topology lacks the subtlety of close reading, coding, and rhetorical analysis, which can better address the "not problem" and other ambiguities that can only be judged in context.

Nonetheless, a computational approach provides some advantages. For one, it allows us to draw on a much broader range of texts than one could analyze using coding and close reading, unless it was possible to conduct such a study with a larger research team. In addition, future iterations of this project will allow us to ask additional questions and form new hypotheses. One extension of this project (currently in progress) involves gathering a second corpus of texts from 2011 to 2015. Using this second corpus, we will be able to track how special and common topics have changed over time. We may hypothesize that some brain regions or concepts will become less prominent in the literature while others will become more prominent as the field progresses.

Relatedly, it may be possible to see if the form of anatomical structures changes over time as techniques for analyzing fMRI data are evolving. There seems to be waning focus on associating regions with functions, and increased focus on functional network connectivity, so we might expect new terms and labels to emerge. In addition, researchers are now using decoding methods like multivoxel pattern analysis that are agnostic to the localization of a function to a region, focusing instead on the pattern of brain activity in a relatively large area. This approach, as well, might require a different set of special topics in order to gain ground, rhetorically.

Another extension of this project will involve tracking the impact factor of individual journals and comparing citation counts for each article to the special and common topics used. This will allow us to determine, for instance, whether some topics are more likely to be cited than others in subsequent studies. Finally, we plan to compare common and special topics to the impact factor of the journal. This will allow us to determine whether certain topics are more likely to appear in higher impact journals, or whether certain rhetorical moves (e.g., employing more or fewer hedges and boosters) correlates with a higher impact factor. These extensions will allow us to identify how, in Derek G. Ross's terms, how neuroscience topics "shape an audience's perceptions and influence decision making because of the way they catalyze frames of understanding."[33] By including citation counts and impact factors in our analysis, we will be able to identify which topics seem most influential and form hypotheses for why this may be the case.

These results will give us a better sense of the power dynamics within the discipline, but additional data will also contribute to this endeavor. For one, funding likely influences the topics researchers study and the approaches they use. Incorporating information about the funding agencies supporting the research for each study would provide a different data point; it may be possible to ascertain, for instance, which topics are most often (and least often) supported by the National Science Foundation or the National Institutes of Health as opposed to private funding sources.

In addition, creating networks based on authors would provide yet another view of the field, one that might demonstrate how key individuals, collaborative teams, or institutions drive developments in the field. Given that neuroscience research, especially research using fMRI and other visualization equipment, is expensive, external funding and institutional support become important resources. The most successful authors and collaborative teams are likely to be those who also have found ways to access those resources.

These data points, together, will help us to develop maps that more clearly demonstrate vectors of power in cognitive neuroscience. Based on citation data, author networks, and funding information, we might be able to determine how specific *topoi* are driving research in cognitive neuroscience. For example, one hypothesis to test is whether authors, teams, citations, and funding constellate around prominent neurological conditions (such as autism or Alzheimer's disease) that are deemed key public health issues, around topics that seem most likely to garner publicity in popular discourse (such as sex differences) simply for their appeal to audiences, or around more basic neurological processes that reflect research specializations more so than external concerns.

Our approach, then, helps to answer Gross, Harmon, and Reidy's call for a "second generation" methodological program, which should include principles for "sampling the scientific texts that will be the object of analysis" and some system for "accounting for change [in rhetorical practices] over time."[34] Moreover, our approach allows us to identify relations of power within the field, as we can see how some *topoi* gain influence over time as others fall away.

Notes

1. Shelly Fan, "Scientists Complete the Most Detailed Map of the Brain Ever," *SingularityHub*, July 31, 2016, http://singularityhub.com/2016/07/31/scientists-complete-the-most-detailed-map-of-the-brain-ever.
2. Matthew F. Glasser et al., "A Multi-modal Parcellation of Human Cerebral Cortex," *Nature* 536, no. 7615 (2016): 171–178.
3. Carolyn R. Miller, "The Aristotelian *Topos*: Hunting for Novelty," in *Rereading Aristotle's Rhetoric*, eds. Alan G. Gross and Arthur E. Walzer (Carbondale: Southern Illinois University Press, 2000): 142.
4. Richard C. Huseman, "Modern Approaches to the Aristotelian Concept of the Special Topic," *Central States Speech Journal* 15, no. 1 (1964): 21.
5. Carolyn R. Miller, "Aristotle's 'Special Topics' in Rhetorical Practice and Pedagogy Author(s)," *Rhetoric Society Quarterly* 17, no. 1 (1987): 67.
6. Lawrence J. Prelli, *A Rhetoric of Science: Inventing Scientific Discourse* (Columbia, SC: University of South Carolina Press, 1989): 199.
7. Lynda Walsh, "The Common *Topoi* of STEM Discourse: An Apologia and Methodological Proposal, With Pilot Survey," *Written Communication* 27, no. 1 (2010): 147, 149.
8. Miller, "The Aristotelian *Topos*," 142.
9. Lynda Walsh and Andrew B. Ross, "The Visual Invention Practices of STEM Researchers: An Exploratory Topology," *Science Communication* 37, no. 1 (2015): 122.
10. Walsh, "The Common *Topoi* of STEM Discourse," 127.
11. For more detailed methodological description, see Elizabeth L. Beam, Gregory Appelbaum, Jordynn Jack, James Moody, and Scott A. Huettel, "Mapping the Semantic Structure of Cognitive Neuroscience," *Journal of Cognitive Neuroscience* 26, no.9 (2014): 1949–1965.

12. Beam et al., 1958.
13. Russell A. Poldrack, "Can Cognitive Processes Be Inferred from Neuroimaging Data?" *TRENDS in Cognitive Sciences* 10, no. 2 (2006): 59.
14. Ibid., 59.
15. Ibid., 63.
16. Nancy Kanwisher, Josh McDermott, and Marvin M. Chun, "The Fusiform Face Area: A Module in Human Extrastriate Cortex Specialized for Face Perception," *The Journal of Neuroscience* 17, no. 11: 4302–4311.
17. B. T. Thomas Yeo et al., "The Organization of the Human Cerebral Cortex Estimated by Intrinsic Functional Connectivity," *Journal of Neurophysiology* 106, no. 3 (2011): 1125–1165.
18. Ken Hyland and Polly Tse, "Metadiscourse in Academic Writing: A Reappraisal," *Applied Linguistics* 25, no. 2 (2004): 157.
19. Ken Hyland, *Metadiscourse: Exploring Interaction in Writing* (London: Continuum, 2005).
20. For a list of reporting verbs, see Joel Bloch, "A Concordance-Based Study of the Use of Reporting Verbs as Rhetorical Devices in Academic Papers," *Journal of Writing Research* 2, no. 2 (2010): 219–244.
21. Walsh, "The Common *Topoi* of STEM Discourse," 134.
22. Ibid., 126.
23. C. A. Boettiger et al., "Immediate Reward Bias in Humans: Fronto-parietal Networks and a Role for the Catechol-O-methyltransferase 158(Val/Val) Genotype." *Journal of Neuroscience* 27, no. 52 (2007): 14383, emphasis added.
24. P.V. Chandrasekhar et al., "Neurobiological regret and rejoice functions for aversive outcomes," *NeuroImage* 39, no. 3 (2008): 1472, emphasis added.
25. Boettiger et al., 14383, emphasis added.
26. Chandrasekhar et al., 1472, emphasis added.
27. See, for instance, David R. Gruber, "Reinventing the Brain, Revising Neurorhetorics: Phenomenological Networks Contesting Neurobiological Interpretations," *Rhetoric Review* 35, no. 3 (2016): 239–253; and Jenell Johnson and Melissa Littlefield, "Lost and Found in Translation: Popular Neuroscience in the Emerging Neurodisciplines" in *Sociological Reflections on the Neurosciences* (Wagon Lane, UK: Emerald Group Publishing Limited, 2011): 279–297.
28. David R. Gruber, "The Extent of Engagement, the Means of Invention: Measuring Debate about Mirror Neurons in the Humanities and Social Sciences," *Journal of Science Communication* 15, no. 2 (2016): 1–17.
29. Clement Levallois, John A. Clithero, Paul Wouters, Ale Smidts and Scott A. Huettel, "Translating Upwards: Linking the Neural and Social Sciences via Neuroeconomics," *Nature Reviews Neuroscience* 13 (2012): 789–797.
30. As an example of topic modeling, see Russell A. Poldrack et al., "Discovering Relations Between Mind, Brain, and Mental Disorders Using Topic Mapping," *PLoS Computational Biology* 8, no. 10 (2012): 1–14.

31. How tasks relate to the cognitive phenomena they study has been a focus of the Cognitive Atlas. See Russell A. Poldrack et al., "The Cognitive Atlas: Toward a Knowledge Foundation for Cognitive Neuroscience," *Frontiers in Neuroscience* 5 (2011): 1–3. (http://journal.frontiersin.org/article/10.3389/fninf.2011.00017/full).
32. S.J. Heinen et al., "An Oculomotor Decision Process Revealed by Functional Magnetic Resonance Imaging," *Journal of Neuroscience* 26, no. 52 (2006): 13515.
33. Derek G. Ross, "Common Topics and Commonplaces of Environmental Rhetoric," *Written Communication* 30, no. 1 (2013): 96.
34. Alan G. Gross, Joseph E. Harmon, and Michael S. Reidy, *Communicating Science: The Scientific Article from the 17th Century to the Present* (Oxford: Oxford University Press, 2002): 8.

Bibliography

Beam, Elizabeth, Gregory Appelbaum, Jordynn Jack, James Moody, and Scott A. Huettel. 2014. "Mapping the Semantic Structure of Cognitive Neuroscience." *Journal of Cognitive Neuroscience* 26, no. 9: 1949–1965.

Bloch, Joel. 2010. "A Concordance-Based Study of the Use of Reporting Verbs as Rhetorical Devices in Academic Papers." *Journal of Writing Research* 2, no. 2: 219–244.

Boettiger, C.A., et al. 2007. "Immediate Reward Bias in Humans: Fronto-Parietal Networks and a Role for the Catechol-O-methyltransferase 158(Val/Val) Genotype." *Journal of Neuroscience* 27, no. 52: 14383–14391.

Chandrasekhar, P.V., et al. 2008. "Neurobiological Regret and Rejoice Functions for Aversive Outcomes." *NeuroImage* 39, no. 3: 1472–1484.

Fan, Shelly. 2016. "Scientists Complete the Most Detailed Map of the Brain Ever." *SingularityHub*, July 31. http://singularityhub.com/2016/07/31/scientists-complete-the-most-detailed-map-of-the-brain-ever.

Glasser, Matthew F., et al. 2016. "A Multi-Modal Parcellation of Human Cerebral Cortex." *Nature* 536, no. 7615: 171–178.

Gross, Alan G., Joseph E. Harmon, and Michael S. Reidy. 2002. *Communicating Science: The Scientific Article from the 17th Century to the Present.* Oxford: Oxford University Press.

Gruber, David R. 2016a. "Reinventing the Brain, Revising Neurorhetorics: Phenomenological Networks Contesting Neurobiological Interpretations." *Rhetoric Review* 35, no. 3: 239–253.

———. 2016b. "The Extent of Engagement, the Means of Invention: Measuring Debate About Mirror Neurons in the Humanities and Social Sciences." *Journal of Science Communication* 15, no. 2: 1–17.

Heinen, S.J., et al. 2006. "An Oculomotor Decision Process Revealed by Function-al Magnetic Resonance Imaging." *Journal of Neuroscience* 26, no. 52: 13515–13522.

Huseman, Richard C. 1964. "Modern Approaches to the Aristotelian Concept of the Special Topic." *Central States Speech Journal* 15, no. 1: 21–26.

Hyland, Ken. 2005. *Metadiscourse: Exploring Interaction in Writing*. London: Continuum.

Hyland, Ken, and Polly Tse. 2004. "Metadiscourse in Academic Writing: A Reappraisal." *Applied Linguistics* 25, no. 2: 156–177.

Johnson, Jenell and Melissa Littlefield. 2011. "Lost and Found in Translation: Popular Neuroscience in the Emerging Neurodisciplines." In *Sociological Reflections on the Neurosciences*, ed. Martyn Pickersgill and Ira Van Keulen. 279–297. Wagon Lane: Emerald Group Publishing Limited.

Kanwisher, Nancy, Josh McDermott, and Marvin M. Chun. "The Fusiform Face Area: A Module in Human Extrastriate Cortex Specialized for Face Perception." *The Journal of Neuroscience* 17, no. 11: 4302–4311.

Levallois, Clement, John A. Clithero, Paul Wouters, Ale Smidts, and Scott A. Huettel. 2012. "Translating Upwards: Linking the Neural and Social Sciences Via Neuroeconomics." *Nature Reviews Neuroscience* 13: 789–797.

Miller, Carolyn R. 1987. "Aristotle's 'Special Topics' in Rhetorical Practice and Pedagogy Author(s)." *Rhetoric Society Quarterly* 17, no. 1: 61–70.

———. 2000. "The Aristotelian *Topos*: Hunting for Novelty." In *Rereading Aristotle's Rhetoric*, ed. Alan G. Gross and Arthur E. Walzer. Carbondale: Southern Illinois University Press.

Poldrack, Russell A. 2006. "Can Cognitive Processes Be Inferred from Neuroimaging Data?" *TRENDS in Cognitive Sciences* 10, no. 2: 59–63.

Poldrack, Russel A., et al. 2011. "The Cognitive Atlas: Toward a Knowledge Foundation for Cognitive Neuroscience." *Frontiers in Neuroscience* 5: 1–3.

———. 2012. "Discovering Relations Between Mind, Brain, and Mental Disorders Using Topic Mapping." *PLoS Computational Biology* 8, no. 10: 1–14.

Prelli, Lawrence J. 1989. *A Rhetoric of Science: Inventing Scientific Discourse*. Columbia: University of South Carolina Press.

Ross, Derek G. 2013. "Common Topics and Commonplaces of Environmental Rhetoric." *Written Communication* 30, no. 1: 91–131.

Walsh, Lynda. 2010. "The Common *Topoi* of STEM Discourse: An Apologia and Methodological Proposal, with Pilot Survey." *Written Communication* 27, no. 1: 120–156.

Walsh, Lynda, and Andrew B. Ross. 2015. "The Visual Invention Practices of STEM Researchers: An Exploratory Topology." *Science Communication* 37, no. 1: 118–139.

Yeo, B.T. Thomas, et al. 2011. "The Organization of the Human Cerebral Cortex Estimated by Intrinsic Functional Connectivity." *Journal of Neurophysiology* 106, no. 3: 1125–1165.

CHAPTER 8

Topology and Psychoanalysis: Rhe-torically Restructuring the Subject

Jake Cowan

Every knot says a lot

—James Joyce

Not for nothing, the final turn of Jacques Lacan's teaching has been overlooked by all but the most fervid of his followers, with barely a word on the subject coming out of rhetoric and composition. That lack is understandable: Only a handful of texts and seminars from this period are currently translated and fewer still readily accessible, whether because those manuscripts remain unpublished or due to the enormous challenge they present. In the muddled transcripts of his last seminars, Lacan's already abstruse prose becomes practically (and purposefully) illegible, littered with tortuous topological diagrams and wordplay that are *stricto sensu* impossible to translate—an especially rich limitation in the context of Lacan's interest in the *impossible* Real. For instance, the penultimate Seminar XXVI of 1978–79, *Topology and Time*, is remembered less for Lacan's typical pomposity than for his "immense weariness, his absences, his silences ... that could last for almost entire meetings," only occasionally "interrupted by a formula written on the board" or by his picking up

J. Cowan (✉)
Department of English, University of Texas at Austin,
Austin, TX, USA

© The Author(s) 2017

L. Walsh, C. Boyle (eds.), *Topologies as Techniques for a Post-Critical Rhetoric*, DOI 10.1007/978-3-319-51268-6_8

some loose string to demonstrate his practice with knots.[1] Despite such difficulties, these later developments make for formative if formidable ground not only in clinical practice, but also as an alternative account of writing. Moving beyond the constraints of ordinary signification that he had identified in his earlier work, toward instead a writing that belongs to the Real, the Lacanian turn toward topology marks the analyst at his most closely aligned with rhetorical theory.

Since the close of the twentieth century, there has been a resurgence of interest in the intersection of psychoanalysis and rhetoric, thanks to the likes of Joshua Gunn, Christian Lundberg, and Thomas Rickert—writers who each offer a unique uptake of (post-)structuralist psychoanalysis. Gunn's articles have focused principally on rethinking rhetoric in terms of Lacan's earliest work on the mirror stage and the related Imaginary register, especially those "psychical-rhetorical narratives" that constitute the "coping fantasy" that a human subject belongs to (yet is distinct from) the surrounding coherent, objective, meaningful world.[2] Both in direct response to Gunn and in other writings, Lundberg alternatively proposes "the possibility of a Lacanian path in rhetorical studies that figures the primacy of the Symbolic as a quintessentially *rhetorical* order,"[3] an order that comes to name a "tropological charge of discourse" that is prior to and formative of the subject's "imaginary commitments" to an agentive, suasive fantasy.[4] The theorist to have given the most central place to the Real as it relates to rhetoric is Rickert, albeit in a form more indebted to Žižek's ideologico-cultural prepossession than to Lacan's clinical interest. Rickert approaches the Real in terms of the "Act," which describes "a particularly potent form of invention, or an unleashing of the evental within ongoing, belated processes of symbolic integration,"[5] thereby reimagining rhetoric beyond persuasive identification toward productive individuation.

While these approaches are valuable, rhetoricians have shied away from the analyst's final developments chiefly because in them, his earlier reliance on linguistics appears to give way to mathematics, from (Symbolic) tropology to (Real) topology. For example, in order to relocate the late Lacan back within the field of rhetoric, Lundberg must dismiss the primacy given to set theory and formal mathematics in the final seminars, claiming that these "appealed to Lacan because [they] afforded him a vocabulary for understanding the subject as an effect of certain repeatable relations," which is to say, mathematization is only "a metaphor that has the benefit of figuring the subject in desubjectivized terms, outside of the conventional vocabularies of intersubjectivity and social construction."[6] As I will demonstrate, however, more than metaphor is at stake for Lacan's final formulations, and his mathematical turn is less a turn away from rhetorical theory than the culmination of his "return to Freud." If anything, Lacan's appropriation of mathematical figures leads away from a metaphorical treatment of the unconscious as structured *like* a language—which would be learnable and communicative—to the Real rhetorical force of language, its productive and affective dimension as born out in writing, which would be indissoluble from the subject of the unconscious. Psychoanalytic topology, then, provides a pathway away from thinking rhetoric in terms of metaphors or narratives that communicate to the outside a speaker's inner world, and toward the production of an always already immanent "*desire*-in-language" that Lacan names *lalangue*, a concept Victor Vitanza has justly linked with other rhetorics of resistance, from "Foucaultian 'counter memory'" to "sophistic *Letteraturizzazione* and Bataille's excess," including the play of Vitanza's own theatricks.[7]

Over the following pages I attempt to chart new ground within rhetoric by tracing the genealogy of psychoanalytic topology from its origins in Freud's topographical heuristics to the tortuous letters produced by Lacan's encounter with the literary avant-garde of James Joyce and Samuel Beckett. Inasmuch as this development constitutes a renunciation of Euclidian logic and coherent grammar in the classical sense, this chapter suggests that Lacanian topology functions as a radical pivot toward a psychoanalytic praxis of performative rhetoric closely aligned with the preoriginary condition of response-ability that Diane Davis has called our *rhetoricity*. While this shift could lead us to rehash the standard squabble over whether psychoanalysis is prescriptive or descriptive, my aim here is instead to draw attention to how topological sets in general, and in particular the nonplanar figures of knot theory, allow critical rhetorician and analyst alike to articulate the various situations and discourses the subject is caught up within without diminishing their multiplicity. As I will show, the Lacanian account provides a way of realizing the overdetermined psychoanalytic subject as topological, rhetorically structured without division between the exterior reality of the (specific) other and the imminent interiority of the (grand) Other. Such a topology rewrites the unconscious as a synchronic and fluid network of letters rather than a diachronic signifying chain, as socially imminent to the subject as the unconscious is personally intimate. If in his turn toward topology Lacan suggests new ways of writing the subject, then this chapter will suggest how the intersection of topology and psychoanalysis indicates new ways for subjects to write—a praxis that that remains open to the Other's discordant diversity, acknowledging a debt of response-ability while seeking fresh sets of rhetorical relationships.

From Two to Three

At least initially, Lacan drew his notion of an imminent and nonplanar psychical *topology* from a source he would greatly reshape: Freud's classical development of different descriptive *topographies*. While these alternative methodologies both relate to manners of rendering place, Freud's is distinguished from Lacan's insofar as topo-*logy*, broadly speaking, "refers to the connectedness of things," while topo-*graphy* is typically "used to denote a discrete place," as with mapping a static locality.[8] Although he had shown interest as early as the 1953 Rome address in the figures of topology—originally coined *analysis situs* by Poincaré, a name which as easily might have covered the (other) scene of the clinic—Lacan's persistent desire for more rigorous ways of writing psychical structure and his growing emphasis on the signifying impossibilities of the Real led him to provide topological illustrations with an increasingly prominent place within his teaching. After first following Freud's vacillations on the subject of the unconscious (especially in its relation to/within spaces of interiority and exteriority), I will clarify how Lacan attempted to solve the same dilemma that plagued his precursor.

Bookending the first few decades of psychoanalysis, during which he was defining the basic contours of the talking cure, Freud wavered between two distinct topographic conceptualizations of the psychical apparatus. The first topography, the one Lacan aimed to recover, stemmed from the earliest descriptions of the unconscious in *Studies on Hysteria* (1895) and its subsequent delineation from the preconscious and conscious systems in *The Interpretation of Dreams* (1900). Yet as Laplanche and Pontalis explain: "this discovery is not limited to the recognition of the existence of distinct psychical locations," for "it further assigns to each of them a separate character and operational mode" within a dynamic and oppositional frame; in other words, "the unconscious itself is organised in strata, and analytic investigation proceeds of necessity via specific paths which presuppose a specific order among the groups of ideas," strata not based on any supposed neurological location, but rather a logico-temporal dynamism.[9] With reference to Freud's 1915 metapsychological essay, *The Unconscious*, Blum and Secor clarify this process succinctly:

> Psychic formations (thoughts, wishes, memories) do not move from one region to another within a mapped and fortified psychic space. What distinguishes a preconscious thought from an unconscious one is not its location but rather its connection to language, the "overinvestment" that happens when an idea is put into words and thus becomes not necessarily conscious but potentially so. The unconscious and the preconscious, then, are not regions, territories, or locations, whether anatomical or structural, but different relationships to the possibility of language.[10]

Hence Freud argued in these early texts against the nineteenth century's predominant anatomical model, which attempted to specify and materialize mental phenomena within the biological brain. This is why Freud early on renounced the more common noun "subconscious" (*Unterbewusstsein*) in favor of the "unconscious" (*Unbewusstes*), for only the latter, "by virtue of the negation that it contains, is able to express the topographical split between two psychical domains and the qualitative distinction between the processes that occur therein."[11] As such, if the unconscious in effect constitutes "an other scene" (*eine andere Schauplatz*—note the importance already placed on *Platz/topos/*place), then it is so only insofar as the unconscious realm is not simply physical, not tangibly locatable as in phrenology; and thus the "topographical mode of representation" outlined in *The Interpretation of Dreams* remains no more than an "expedient and justifiable" heuristic, an "illustrative" metaphor to make intelligible an otherwise "*virtual*" system.[12]

For all his pained efforts to avoid confusing the imaginary map with a corresponding neurobiological domain, Freud eventually abandoned his first topography on account of two inherent barriers: On one level, he was clearly "frustrated by the geometric, material limitations that such a model imposes," unable to escape the linearity and boundaries Euclidean representation requires[13]; simultaneously, Freud recognized that an organization of dynamic strata was too rigid to account for the influence of the external world on a developing personality, such as the penitent conscience that often follows from having a harsh parent. Accordingly, Freud attempted to rectify this impasse through a second topography, also called his "structural model," which originated in 1923s *The Ego and the Id*, laying out the tripartite interrelationship of the titular agencies alongside the superego. While both of these theoretical constructs have enjoyed profound influence throughout the humanities, their conceptual deficiencies present a problem of particular interest to rhetoricians (as will the topological corrective Lacan proposed)—namely, how they broach the pedagogical limits of metaphor and signification, necessitating a reconceived writing praxis that seeks alternatives to fantasies of meaningful communication.

Like the previous topography—which does not easily square with its alternative—the components of Freud's later model are defined dynamically, as functions rather than locatable entities; yet as Laplanche and Pontalis point out, the new terminology is no longer based on a stratified metaphor "borrowed from the physical sciences but … instead shot through with anthropomorphism: the intrasubjective field tends to be conceived of after the fashion of intersubjective relations, and the systems are pictured as relatively autonomous persons-within-the-person."[14] This second map, however, ended up reinscribing many of the same limits as Freud's first topography, mainly because he never entirely gave up his earlier model, even trying at points to merge the two visually.[15] The frontiers dividing the agentive trio were no longer sharply defined, but as their German designations make clear, the "it" (*Es*/id), the "I" (*Ich*/ ego), and especially the "Over-I" (*Überich*/superego) still retained a basic spatial character, differentiated through a locative metaphor. Moreover, in Lacan's critique, the attempts by Freud and others (particularly those in the ego-psychology camp) to harmonize the two models allowed this terminology "to slide from the ego [wrongly] defined as the perception-consciousness system—that is, as the system of the subject's objectifications—to the ego conceived of as the correlate of an absolute reality." In this way, "[o]nce the intrasubjective topography has become entified," there results "a sort of psychological orthopedics" wherein the labor of the clinic becomes "the relation of two bodies between which a fantasmatic communication is established in which the analyst teaches the subject to apprehend himself as an object," a process by which a misrecognized "subject, transformed into an *it*, has to conform to an ego which the analyst has no trouble recognizing as his ally, since it is, in fact, the analyst's own ego."[16] This means that the reorientation toward exterior influences of the second topography—which would not be *structural* in the Lacanian sense, for it only comprises "a theoretical model" and not "the original machine that directs the subject"[17]—does not resolve the former's dilemma, but tends merely to flip emphasis, reproducing a binary between inside and outside, even as the latter now weighs tyrannically on the former.

Against this, and against adaptation of the analysand to the analyst's ego in general, Lacan makes a turn toward topo-*logy* over topo-*graphy* in order, in part, to figure the question of interiority, finding in nonplanar or nonspherical objects a means to realize "the immanent, nontotalized structure of the subject."[18] This shift is methodologically crucial because, for the late Lacan, "[t]opology is not 'made to guide us' in the structure" in the form of yet another metaphorical heuristic, but rather "[t]opology *is* this structure—as a retroaction of the chain order of which consists language"[19]—and thus "goes much further than an intuitive schema" like Freud's topography.[20] According to Jacques-Alain Miller's commentary, there is an "a priori lack of correspondence between a graphic representation and its object ... *in the space of intuition*," such as with imaginary analogs of the Freudian models that "have only a didactic role;" and while this means for the young structuralist of *Écrits* that "*there is no longer any occultation of the symbolic* in the topology that Lacan establishes, because this space is the very space in which the subject's logical relations are schematized,"[21] for the other Lacan of *Autres Écrits*, the effect and emphasis of topology has shifted away from the Symbolic toward a continuous articulation with the impossible Real, that which produces the "occultation of the symbolic" because "[s]tructure *is* the real, which shows itself in language."[22] Inasmuch as in its classical usage "topography was primarily a form of narrative" that Freud reiterated throughout his oeuvre,[23] Lacan's topo-*logical* reworking of Freudian psychoanalysis—which follows the affirmative logic of preservation through continuous deformation—no longer aims for narrative coherence but instead underscores a subject's relationship to the ungrammatical (Imaginary) and alogical (Symbolic) aspects of language: otherwise said, *the subject of psychoanalysis, the subject of the Real, is therefore also the subject of rhetoric.*

As Alfredo Eidelsztein has advanced, there are five "dimensions" to mathematical topology that make it "an appropriate means" by which Lacan is able to write psychoanalytic structure without resorting to imaginary heuristics or reproducing Freud's dualisms.[24] In the first place, "shapes do not perform any function in topology," just as in the clinic, imaginary and intuitive "shapes or forms do not perform a determinant function," remaining at the level of conscious appearances. Secondly, "in topology no measurable function of distance or size is considered," while "[i]n psychoanalysis, we apply these properties to time and space," which likewise do "not work according to a measure." The third dimension is that "topology allows us to work using a new relation between exterior and interior," common categories that, from a Lacanian perspective, are as imaginary as shape and form—a point I will return to shortly. Fourth, "topology turns upside down the usual concept of the object/subject relation," such as the Cartesian binary between three-dimensional *res extensa* and nondimensional *res cogitans* ("knowledge is no load to carry"); alternatively, says Eidelsztein: "thanks to and through the articulation between psychoanalysis and topology, we move from the pair three-dimensionality/non-dimensionality towards the two-dimensional object and subject," no longer set in opposition but conjoined through the conjunction. Finally, once all these commonplace distinctions—shape, measurement, inside/outside, subject/object—have begun to vanish, all that remains are formulaic invariants, and "that is the structure" realized by Lacanian topology, a practice which aims at the fixed relations between elementary yet open, determinative yet undetermined *loci* within covariant sets. By stripping the subject down to a minimal aggregate of indefinite and nonseparate variables and their relations, topology equips the Lacanian analyst (whether of the psyche or of rhetoric; it amounts to the same insofar as the unconscious is structured like a language) with an exceptionally plastic technique for writing subjective and affective transformations beyond narrative or metaphor.

To say, then, that topology presents (rather than metaphorically represents) the Real is to point to a ternary logic that upsets the ordinary conception of structure as the difference between surface and depth, or in Lacanian terms, the Imaginary and the Symbolic. It is to suggest a way of articulating the subject of the "unconscious structured like a language" beyond signification and the signifier, de-signating it instead as immanent, emergent, and utterly interwoven with performative "linguistricks" and rhetorical production. Topology thus helps us to locate what Vitanza has called rhetoric's "wild (screaming) place of the *third*, which turns the *di*lemma into a *tri*lemma," and so enjoins the writing subject to theatrickally try lemmas (in both mathematical and morphological senses), that is, to accept whatever the Other might offer in order to unearth/destratify the disruptive and resistant "place of Babel/babble, the place of the *pagus*, the place where the most 'radical [of] multiplicities' dis/engage in, heretofore unknown, linkages."[25] Moreover, as Dany Nobus has argued, as a response to post-structuralist critiques lobbed at Lacan's "system," the topological turn allowed Lacan to distance his theory from structuralism's inherent presupposition "of a total and totalizing language system centered around the primordial incidence of the signifier"—a drive toward totality echoed by the "blatant Eurocentric rhetoric of discovery"[26] that characterized Freud's topographic constructions. Instead, Lacan's "recourse to topology allowed" the omission of any transformative agency wholly under conscious control.[27] Because they are formed around precisely these sort of central voids, nonspherical figures such as the torus or the Möbius strip provide ways to reconceive the psychoanalytic subject without falling into the representational traps of the Freudian models with their problematic split between inside and outside. What Lacan's nonplanar applications suggest instead is that there exists a continuity between what is within and what is without the subject, that in its inmost being there exists a trace or a kernel of what remains other to the subject: Lacan's neologism for this inner exteriority is "extimacy" (*extimité*), which, in J.A. Miller's explanation, "is not the contrary of intimacy [*intimité*]," but rather "says that the intimate is Other—like a foreign body, a parasite" at the very heart of its host.[28]

By this view, the extimate subject has the structure of a torus: Like the human body, which at one level is a tube processing material only nominally contained "within" our digestive machinery, the Lacanian subject is equivalent to its various openings unto the outside world, simultaneously set between these gaping cuts (mouth, anus, ear, eye, and so on) and cut by the circuit that connects them. (This gives new meaning to Lacan's quip that the signifier's answer to whoever questions it is to "[e]at your Dasein":[29] Not only does it mean that the subject of the signifier must confront her finitude in the face of desire's Symbolic perpetuation, but it also means that the toric subject must learn to live "excentrically," finding the kernel of her being *there* in the parasitical, extimate space she does not occupy, yet which occupies her.) At the same time, because the difference between *this* (in)*side* and *that* (out)*side* can only be determined diachronically, appearing synchronically as a one-sided infinite loop, the subject's structure is that of the Möbius strip—which, and this is the (w)hole point, can be transformed into a torus by doubling the surface (combining two individual bands along one shared edge) and adding empty volume to the new interior,[30] a lack in the shape of a nonorientable surface, as in the split tori sculptures of Keizo Ushio (Fig. 8.1).[31]

Being continuous does not mean, however, that the subject and the other, inside and outside, are unified or equivalent for Lacan; rather, the *(a)crucial* lesson of the Möbius strip is how nonspherical topology implies neither a homogenous unity nor an opposing duality, but a wavering triplicity: While traversing these figures reveals their oneness, there nonetheless remains alternating sides at any given point taken by itself, so that there exists both two and three at one and the same time.[32] In *The Unnamable* (another way to say "the Real"), Beckett came closest to putting this form of subjectivity into words when he described the self as:

Fig. 8.1 Keizo Ushio, *Oushi Zokei*, 2010. By User: Moondyne/ Wikimedia Commons/ CC BY-SA 3.0 (cropped from original)

> an outside and an inside and me in the middle, perhaps that's what I am, the thing that divides the world in two, on the one side the outside, on the other the inside, that can be as thin as foil [*in literary terms, its own opposite*], I'm neither one side nor the other, I'm in the middle, I'm the partition, I've two surfaces and no thickness ... on the one hand the mind, on the other the world, I don't belong to either....[33]

Topology thereby comes to provide Lacan with a solution to the question of interiority that plagued Freud, not by identifying the psychoanalytic subject with one side or the other, but by *showing* in the use of impossible figures how the two alternatives are coterminous and how the subject belongs to both and n(one). Lacan thus links the realization of a subject to a writing praxis that entails dealing with time and place, a rhetoric that must be materially traversed and performed. Moreover, as the following section will elaborate, this provided a path for Lacan to return to the original emphasis on the unconscious as psychoanalysis' central revolution, an "unconscious structured like a language" which, in light of Lacan's late developments, may be reformulated as *an unconscious topologized like a-language*.

The Deal with the Real

Thus far I have outlined why Lacan turned to the field of formal topology in his later teaching, contending that Lacan formulated his topology as a response and a return to Freud's classical topographic models, particularly the problem of interiority that had led, in Lacan's view, to the ethical quagmire of Anglo-American ego-psychology. While I have up till this point attempted to focus primarily on this short genealogy and what topology means for the psychoanalytic subject, this limited approach has inevitably led to questions and references that now call for closer attention. Specifically, the registers of the Real, Symbolic, and Imaginary (RSI) must be reconsidered in light of another topological figure developed in detail by Lacan, the Borromean knot—for if structure is Real and topology is structure, then the RSI trilemma's co-inherence must be reinterrogated and rewritten. As I hope to show, such rewriting is especially important for an understanding of how psychoanalytic topology shapes language beyond the confines of the signifier and why the toric subject can also be considered the excentric subject of rhetoric.

In its more familiar formulation, which Lacan began to develop in 1953 around the same time as his first topological references, the RSI trifunctionality names the three orders (or registers) of psychical experience that define, through their contingent interrelationship, the specific psychopathology of the clinical subject, and without which "it would be impossible to understand anything of the Freudian technique and experience."[34] Despite this practical charge, however, the Lacanian orders do not correspond to either of the Freudian topographies. In short and simplified fashion, *the Imaginary*, which preoccupied the earliest Lacan, correlates to experiences of autonomy and coherence, the belief that things really are just as they (literally) appear to be, that reality is by and large whole and trustworthy. Its paradigmatic instance, perpetually repeated in daily life and not just infancy, is the mirror stage, when a child (mis)recognizes (*méconnaît*) itself in/as its reflection; as such, the ego belongs to the Imaginary register (as does ego-psychology, hence Lacan's constant ire), and so too relationships of similarity, identification, and binary balance (such as the inside/outside dichotomy already discussed). *The Symbolic*, on the contrary, refers not to the world of self-same appearances but to the underlying dimension of difference and language, which is why it governed Lacan's thought throughout his structuralist period, when he defined the role of analysts as "practitioners of the symbolic function."[35]

Whereas in the Imaginary there occurs a (mis)identification between the self (subject) and its reflection (ego), the Symbolic is characterized by the failure of any fixed relation whatsoever, being instead the dominion of radical alterity that Lacan will often refer to as the big Other, which keeps open the differential chain as it mediates and triangulates the subject. Insofar as language, following Saussure, is formed through a system of interminable otherness wherein a word gains its specificity according to what it is not, the Symbolic has a close association with our means of communicative exchange, and in particular the function of the signifie-*r*; the signifie-*d*, on the other hand, which suggests a stable reference, belongs to the Imaginary, and it is through the overlap of the two registers where meaning emerges. Yet to remain at this duality between Symbolic signifier and Imaginary signified would suggest the subject is stuck in the Imaginary, which is the order of binaries, hence a third is needed to distinguish the subject. For this and for other reasons, because not all (*pas tout*) can be contained within the confines of linguistic structure, since *not all can be said*, Lacan posits that something further must *ex-ist* outside meaning, beyond signification, and he calls this (a)logically necessary register *the Real.*

Although he had made reference to this last order early in his oeuvre, the Real does not take center stage until the late 1960s, and by that point, the category had developed past both the commonplace definition of "what resists symbolization absolutely"[36] and the base materialism sometimes associated with it. Reconceived topologically, the Real is loosed from such Kantian connotations (of the noumenal *Ding*) and is described instead as the impossible, as what cannot be accounted for in the economic machinery of the Symbolic; more than an irreducible remainder, the Real "admits something new, which is precisely the impossible."[37] To this formulation, Lacan adds in the early 1970s that the impossible is what "doesn't stop not being written," and subsequently, in the same seminar, he begins using the Borromean knot to explain the interdependence of psychical trifunctionality—a figure consisting of three rings tied so that if one comes undone, as happens in psychosis, the other two fall apart (Fig. 8.2).[38]

It is important to underscore here the Real is absolutely not some (imaginary) ideal Thing out there in reality that we only approach asymptotically, which would imply a geometric and not a topological structure; Lacan is especially careful to draw a distinction between the Real and reality, the latter being "merely the fantasy through which thought sustains itself—'reality' no doubt, but to be understood as a grimace of the real."[39]

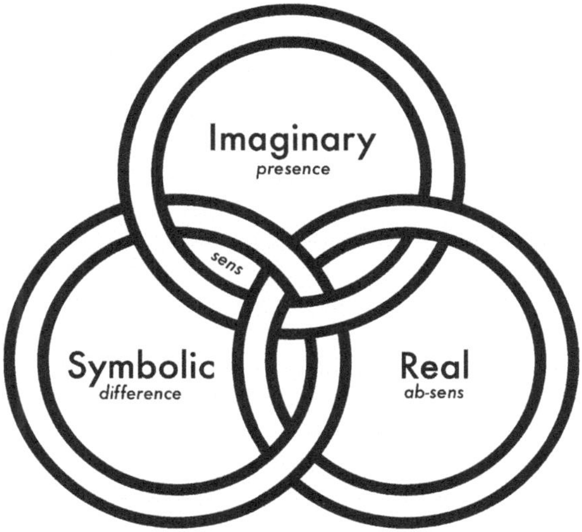

Fig. 8.2 Lacan's Borromean Knot (Commons.wikimedia.org)

Far from this, the Real is a modal category emerging through tension with symbolization, so that when the Real *does* enter language, as it can in avant-garde transgressions, it *no longer simply counts as Real*—though with every new saying, something else goes unsaid and becomes impossible. Neither what is supposedly *really present* (which the Imaginary would capture) nor a matter of merely (Symbolic) *nonpresence*, the Real must be thought of radically existing outside this dichotomy, positing an impossible and unidentifiable third term that ousts meaning in what Lacan calls *ab-sens*[40]: Not simply nonsense but more radically the negation of sense.

What Lacan's neologism aims at is the play of incommensurability and equivocation, of condensation and homophony, the babbling flow of a primordial and polysemous language specific to each subject that breaks through and up any sense of meaningful coherence in a given discourse. Another name he uses to identify the realm of signification ex-collusion is *lalangue*, which is effectively *a-language*; Graciela Prieto gives its etymology, which collapses the article with its significate, as "a reference to the term 'lallation,' i.e. the bath of sounds, both heard and emitted, that the child is immersed in before he acquires articulated language,"[41] and which thereby constitutes the chaotic substrate of the unconscious. Vitanza identifies this flow, unfathomable to cartopographic representations, with the incorporation of the Other into the self, a preoriginary embodiment of Babel through "wild/savage baby-lonialisms, interruptions/asyndetons" that "will have de*term*ined our new (grotesque) parasubjectivities."[42] In the rapturous ruptures of *lalangue*, the Real rips through the Symbolic and unravels the Imaginary, revealing a noncommunicative, performative dimension that undergirds language, which is defined now as no more than "knowledge's harebrained lucubration [*élucubration*] about [*lalangue*]."[43] This *ab-sens*, moreover, gives rise to a *jouissance* (an overwhelming enjoyment beyond pleasure) outside the confines of meaning: A *jouissance* that is not a *jouis-sens* but instead an affective enjoyment of language-in-itself, "*jouissance* located in the jargon itself,"[44] which amounts to an ex-centric enjoyment of a subject's idiosyncratic, idiomatic, idiotic unconscious.

Of note here for rhetoricians is that, as opposed to Symbolic language, "whose chains give rise to the signified as an effect" and hence would fall into the domain of linguistics, *lalangue* as Real and affective "is made up of a multiplicity of elements that convey no meaning in particular, and that are merely the non-sufficient condition of meaning, each being able to receive a plethora of meanings depending on the different linguistic constructions in which they are used."[45] Lacan refers to such an element of *ab-sens*, polymorphously productive yet peripheral to meaning, as *the letter*, and based on his appropriation with the arithmetical figures of topology, we can see how *lalangue* constitutes a covariant set of letters, the Real structure of the subject of the unconscious. The Lacanian letter stands out in the register of the Real as "the material medium [*support*] that concrete discourse borrows from language"[46]—that is to say, through a retroactive effect produced by the Symbolic, the letter is real-ized as "the essentially localized structure of the signifier."[47] An arbitrary mark, the protean shape of which is meaningless prior to annexation by an alphabet, the letter comes to isolate through writing the re-markable unary trait of a signifier, ex post facto transforming the letter into "the basic support for the phoneticization of language," so that "it is speech itself that ultimately finds its [extimate] basis in the exteriority of the written mark."[48] Tzahi Weiss glosses Lacan's looping logic summarily: "The signifier is prior to the signified, and at the foundation of the signifier lies that Real element which does not participate in the sign, which is not assembled from a signifier and a signified, which cannot be interpreted," leading us to the letter.[49]

Here is one of Lacan's great ironies, developed topologically on the model of the torus: The (alphabetic) letter, that which stands at/as the foundation of Western history, that which allows us to communicate across time and distance—the Real of the topological unconscious knows nothing of these categories, yet is nothing without them—is fundamentally unreadable, marked by *ab-sens*, prior to its assimilation within the signifying empire it functionally founds. Hence the trauma of the uninterpretable letter when encountered out of a Symbolic context, as with the claps of Joycean thunder scattered across *Finnegans Wake*: "bababadalgharaghtakamminarronnkonnbronntonnerronntuonnthunntrovarrhounawnskawntoohoohoordenenthurnuk!" In this muddled mouthful, which threads together multilingual signifiers for "thunder," Joyce deals in the letter's meaningless materiality, deforming seemingly natural language back into the primordial polyglot soup of *lalangue*. Of course, not every example needs be this opaque: Beckett's patterns of negation through repetition or Vladimir Nabokov's elegantly elastic alliteration or those sentences of David Foster Wallace that seem to infinitely quest for novel ways to fork or fold back in on themselves—these different techniques likewise render the Real *ab-sens* of the letter. They play with the topological affectability of signification's slippery precondition, thereby disclosing the *seemy* underbelly of language beyond meaning that consciously belongs to nobody (no matter how hard one tries, a plosive and a sibilant will never acoustically resemble one another) even as a body unconsciously belongs in a way to *ab-sens*.

Thus, the architectonic letter both grounds discourse and disturbs it. And while the failure inherent to the signifier produces desire, the agency of the letter produces "a dissident jouissance," as Colette Soler suggests: "It is not so much that jouissance is the referent of the letter, it is rather that the letter is an element of language that is enjoyed."[50] Neither signifier nor signified, the letter is thus Lacan's de-signation for the Real performative force that undergirds language, the elementary makeup of *lalangue* that is affective (*jouissance*-laden) rather than knowably communicative, a rhetoric motivated by and productive of unconscious enjoyment that "must be grasped in the effect that it has and not in its signification."[51] Furthermore, the subject's relation to the topology of its own unconscious alphabet is extimate, in the manner of a torus: The Other at the heart of the self, continuous with the self, is the unconscious topologized by the letter, which is *a-language* invariably beyond and before the subject's knowledge and control; mirroring the Möbius strip, the toric subject becomes essentially indistinguishable from the ex-centric Other of affective *lalangue* that inaugurates its being and might therefore be rewritten as the rhe-torical subject.

Emphasizing the hole within the whole, Lacanian topology thereby comes to reveal an alterity at the heart of "our" singularity, in line (albeit with twists) with what Diane Davis has described as "an originary (or preoriginary) rhetoricity—an affect*ability* or persuad*ability*—that is the condition for symbolic [or Symbolic] action."[52] As "preoriginary," rhetoricity articulates the condition of possibility for (Imaginary) self and (Symbolic) other to emerge: The other does not split the subject as a foreign object but is extimate within its inmost being. Rhetoricity thus exceeds and precedes communicative interpretation, "deal[ing] not in signified meanings but in the address itself, in the exposure [*ex-posure*, "place out"] to the other (*autrui*); it deals not in the said (*le dit*) but in the saying (*le dire*)," the latter, like the letter, "indicat[ing] a nonreferential performative intrusion, an address that necessarily unsettles what is congealed in the already said."[53] Figures such as the torus and the Möbius strip perform the topological structure of rhetoricity *literally*, pointing to the rhe-torical subject's perpetual engagement with the impossibility of *lalangue*.

From a psychoanalytic point of view, "[t]o use language without saying anything is a performance" wherein "meaning flashes from everywhere, every word, every syllable," so that "it is the reader who should decide about meaning," as with the unreadable strings in *Finnegans Wake*.[54] A writing of *ab-sens* founded on the arbitrary letter is thus a rhetorical *doing* shared across the spectrum of reader and/as writer, self and/as other joined in a mutual response-ability as if through the Möbius strip; the Real does not give itself over to intuitive understanding but enjoins us to work through its contortions by putting a letter into context and learning to make do with its otherwise perturbing ambiguity. According to Roberto Harari, what Lacan indicates by his late, so-called silent seminars—when he preferred drawings and drawstrings to further disquisition—is that writing topology, particularly that of knots, "consists, incontrovertibly, in the act of tying it: *it must be tied* [*il faut le faire*], to be understood in a literal sense," which suggests that writing requires a certain "know-how-with [*savior-y-faire-avec*]" to put the Real into practice.[55] With this dense composite, Lacan aims past knowledge as conscious of a referent (*savior*) or as an unconscious dexterity (*savior-faire*) toward instead a nonconceptual knowledge that exists only insofar as it is put into practice, a knowledge that necessitates "sorting out" (*y-faire*) and "dealing with" the contingent Real; accordingly, this "know-how-with" cannot be *known* so much as *done*, consists not in *dit-mention* of the said but *dire-ction* of a saying.[56] It was to develop this *savior-y-faire-avec* that Lacan urged his students to move beyond heuristic schemas in order "to construct his knots with actual strings and to manipulate them, in order to escape the imaginary capture, which would make them lose sight of the Real of their structure."[57] Consider once more the Möbius strip, the continuous unicity of which only becomes apparent when the shape of the figure is methodically traced; similarly, at a level of still more elaborate labor, one reason that the Borromean knot is so difficult to sketch from memory or evaluate at a glance is because its convoluted crossings refuse intuitive apprehension.

Thinking topologically therefore "completely changes the meaning of writing," says Lacan, in that it deals with letters rather than signifiers; hence he redefines writing as "a doing which gives support to thinking," as the Real "support for thinking, for a-thinking (*appensé*),"[58] this latter term naming "a kind of mind control" constituted by the Symbolic chain.[59] "Writing becomes volumetric" when it is approached rhe-torically and performatively, Prieto explains, and that "is true even for the flat scheme where we mark the over-and-under movements of the string. Because they are made of actual pieces of string, these knots are subject to continual twisting," which necessarily "involves a fourth dimension," the temporality of a practiced *savior-y-faire-avec*, "whose imperceptibility to those that inhabit it makes its ex-sistence no less real."[60] Moreover, according to Harari, "the *doing* of something is a kind of vaccination against Imaginary thinking" and the agentive ego, and in defiance of such misrecognition, writing as an imminent praxis, as a *savior-y-faire-avec*, enjoins the subject to encounter the Real, to confront the condition of our rhe-toricity and response-ability to/for the other.[61] Subverting the end Quintilian gave to rhetoric, Prieto suggests that writing rhe-torically, or what she terms "nodal writing," "tries to describe what the stumbling of the *bene dicere* [speaking well] may be able to identify regarding the real of *lalangue*, what may be said about what escapes language"; said otherwise, it underscores the extimate core under signifying language, and "thanks to the half-said [*mi-dire*] truth" of the letter put into practice, writing "can engender new kinds of knowledge" through rhe-torical production,[62] a knowledge of how to make, of how to do, and of how to make do with what we have.

Akin to the Möbius strip it traces, Lacan's topological turn toward the Real moves in two directions at the same time, constituting both a return to and revision of Freud's topographical insights, and proposes an ethico-aesthetic end to analysis—one that, moreover, bears a striking resemblance to that of classical rhetoric. "This is a surprising transformation," says Harari, in that "this notion of *dealing with it*," of developing a *savior-y-faire-avec* the obstinately obscure Real, "comes to change the theory of the end of analysis so that the latter [or the letter] is no longer associated with a search for truth" like an archeological pursuit through the unconscious. Analysis, rather, aims toward discovering a variety of truths beyond the imaginary self-same, toward verity as variety: "This *varity* [*varité*] is thus posited at the expense of *the* truth—as an inflection of the necessary," which is ultimately our response-able dependence on the other, that which we cannot do without, our very rhetoricity.[63] In this way, Lacan comes to replace "the commonplace therapeutic ethics of well-being with a new psychoanalytic ethics of well-saying (*bien-dire*)," along with "the ascendency of a type of speech that is liberated from the s(t)rains of meaning,"[64] using topology to transform the analytic act into a rhe-torical praxis. Yet in place of a traditional emphasis on eloquence, Lacan's suggestion is that when we learn to write with letters and not signifiers, in terms of the Real and not only Symbolic/Imaginary meaning, we end up cultivating *alloquence*: We come to speak and write from out the extimate place of the other (*allos*), attending to an unmasterable *ab-sens* and the error-prone poetry of *lalangue*. In that way, rhe-toric in its performative and a-signifying dimension becomes the end of analysis, while writing topologically becomes "a perpetual expression of desire-as-cure" rather than a cure for our unwieldy, unyielding desires.[65] Written no longer as a holistic sphere, but alternatively as a torus shaped around a black hole, as a Möbius strip for which there is no outside, as a four-dimensional Borromean knot that calls for our *savior-y-faire-avec*, we can see how topology leads Lacan from the "symbolic universe"[66] of his early seminars to declaring in his final ones that the "universe is a flower of rhetoric."[67]

Notes

1. Marcelle Marini, *Jacques Lacan: The French Context*, trans. Anne Tomiche (New Brunswick, NJ: Rutgers University Press, 1992), 247.
2. Joshua Gunn, "Refitting Fantasy: Psychoanalysis, Subjectivity, and Talking to the Dead," *Quarterly Journal of Speech* 90, no. 1 (2004): 2–5, doi:10.1080/0033563042000206808.
3. Christian Lundberg, "The Royal Road Not Taken: Joshua Gunn's 'Refitting Fantasy: Psychoanalysis, Subjectivity and Talking to the Dead' and Lacan's Symbolic Order," *Quarterly Journal of Speech* 90, no. 4 (2004): 496.
4. Christian O. Lundberg, *Lacan in Public: Psychoanalysis and the Science of Rhetoric*, Rhetoric, Culture, and Social Critique (Tuscaloosa: University of Alabama Press, 2012), 25.

5. Thomas J. Rickert, *Acts of Enjoyment: Rhetoric, Žižek, and the Return of the Subject*, Pittsburgh Series in Composition, Literacy, and Culture (Pittsburgh, Pa: University of Pittsburgh Press, 2007), 31 f.
6. Lundberg, *Lacan in Public*, 16.
7. Victor J. Vitanza, *Negation, Subjectivity, and the History of Rhetoric* (Albany: State University of New York Press, 1997), 324.
8. Tim Cresswell, *Geographic Thought: A Critical Introduction* (Malden, MA: Wiley-Blackwell, 2013), 218.
9. Jean Laplanche and Jean-Bertrand Pontalis, *The Language of Psychoanalysis*, trans. Donald Nicholson-Smith (London: Karnac Books, 2006), 450.
10. Virginia Blum and Anna Secor, "Psychotopologies: Closing the Circuit between Psychic and Material Space," *Environment and Planning D: Society and Space* 29 (2011): 1033, doi:10.1068/d11910.
11. Laplanche and Pontalis, *The Language of Psychoanalysis*, 431.
12. Sigmund Freud, *The Interpretation of Dreams*, trans. A. A Brill (New York: Modern Library, 1994), 461.
13. Blum and Secor, "Psychotopologies," 1032.
14. Laplanche and Pontalis, *The Language of Psychoanalysis*, 452.
15. See: Sigmund Freud, *The Ego and the Id*, trans. Joan Riviere (New York: W. W. Norton & Company, 1989), 18.
16. Jacques Lacan, *Écrits: The First Complete Edition in English*, trans. Bruce Fink (New York: W.W. Norton & Co, 2006), 251.
17. Ibid., 544.
18. Blum and Secor, "Psychotopologies," 1035.
19. Jacques Lacan, "*L'Étourdit*," in *Autres Écrits* (Paris: Seuil, 2001), 483, quoted in Samo Tomšič, "Homology: Marx and Lacan," *S: Journal of the Jan van Eyck Circle for Lacanian Ideology Critique* 5 (2012): 112, note 17, emphasis added.
20. Lacan, *Écrits*, 734.
21. Jacques-Alain Miller, "Commentary on the Graphs," in *Écrits: The First Complete Edition in English*, by Jacques Lacan, trans. Bruce Fink (New York: W. W. Norton & Co, 2006), 903.
22. Tomšič, "Homology: Marx and Lacan," 112, note 17, emphasis added.
23. Diane O'Donoghue, "Mapping the Unconscious: Freud's 'Topographic' Constructions," *Visual Resources* 23, no. 1–2 (2007): 106.
24. Alfredo Eidelsztein, *The Graph of Desire: Using the Work of Jacques Lacan*, trans. Florencia F.C. Shanahan (London: Karnac Books, 2009), 4–7.
25. Vitanza, *Negation, Subjectivity, and the History of Rhetoric*, 322, emphasis in the original; the nested citation refers to Jean-Jacques Lecercle's *Philosophy Through the Looking-Glass* (La Salle, IL: Open Court, 1985).
26. O'Donoghue, "Mapping the Unconscious: Freud's 'Topographic' Constructions," 111.

27. Dany Nobus, "Lacan's Science of the Subject: Between Linguistics and Toplogy," in *The Cambridge Companion to Lacan*, ed. Jean-Michel Rabaté (Cambridge: Cambridge University Press, 2003), 64.
28. Miller, Jacques-Alain, "Extimité," in *Lacanian Theory of Discourse: Subject, Structure, and Society*, ed. Mark Bracher et al. (New York: New York University Press, 1994), 76.
29. Lacan, *Écrits*, 29.
30. Nathalie Charraud, "Topology: The Möbius Strip between Torus and Cross-Cap," in *A Compendium of Lacanian Terms*, ed. Huguette Glowinski, Zita M. Marks, and Sara Murphy, trans. Oliver Feltham and Dominique Hecq (London: Free Association Books, 2001), 205.
31. N.A. Friedman and C.H. Séquin, "Keizo Ushio's Sculptures, Split Tori and Möbius Bands," *Journal of Mathematics and Art* 1, no. 1 (2007): 47–57, doi:10.1080/17513470701217217.
32. Steve Pile, "Beastly Minds: A Topological Twist in the Rethinking of the Human in Nonhuman Geographies Using Two of Freud's Case Studies, Emmy von N. and the Wolfman," *Transactions of the Institute of British Geographers* 39 (2014): 232, doi:10.1111/tran.12017.
33. Samuel Beckett, *Three Novels: Molloy, Malone Dies, The Unnamable* (New York: Grove Press, 1991), 383; on Beckett as the paramount writer of the Lacanian letter, see: Suzanne Dow, "Lacan with Beckett," *Nottingham French Studies* 53, no. 1 (2014): 1–18.
34. Jacques Lacan, *Freud's Papers on Technique, 1953–1954*, trans. John Forrester, The Seminar of Jacques Lacan, bk 1 (New York: Norton, 1991), 73.
35. Lacan, *Écrits*, 235.
36. Lacan, *Freud's Papers on Technique, 1953–1954*, 66.
37. Jacques Lacan, *The Four Fundamental Concepts of Psychoanalysis*, trans. Alan Sheridan, Reiss, The Seminar of Jaques Lacan, Book 11 (New York: Norton, 1998), 167.
38. Jacques Lacan, *On Feminine Sexuality: The Limits of Love and Knowledge*, trans. Bruce Fink, The Seminar of Jacques Lacan, bk. 20 (New York: Norton, 1999), 94.
39. Jacques Lacan, "Television," trans. Denis Hollier, Rosalind Krauss, and Annette Michelson, *October* 40 (1987): 10.
40. Jacques Lacan, "L'étourdit," trans. Cormac Gallagher, *The Letter* 41 (2009): 38.
41. Graciela Prieto, "Writing the Subject's Knot," trans. Kristina Valendinova, *Recherches En Psychanalyse* 12 (December 2011): 173, doi:10.3917/rep.012.0170.
42. Vitanza, *Negation, Subjectivity, and the History of Rhetoric*, 52, emphasis in the original.
43. Lacan, *On Feminine Sexuality*, 139.

44. Carmela Levy-Stokes, "*Jouissance*," in *A Compendium of Lacanian Terms*, ed. Huguette Glowinski, Zita M. Marks, and Sara Murphy (London: Free Association Books, 2001), 106.
45. Colette Soler, *Lacanian Affects: The Function of Affect in Lacan's Work*, trans. Bruce Fink (London: Routledge, 2016), 43.
46. Lacan, *Écrits*, 413.
47. Ibid., 418.
48. Aaron Schuster, *Commentary on Lacan Seminar IX L'Identification, 20 December 1961* (unpublished), quoted in Lorenzo Chiesa, "Count-As-One, Forming-Into-One, Unary Trait, S1," *Cosmos and History: The Journal of Natural and Social Philosophy* 2, no. 1–2 (2006): 79.
49. Tzahi Weiss, "On the Matter of Language: The Creation of the World from Letters and Jacques Lacan's Perception of Letters as Real," *Journal of Jewish Thought & Philosophy* 17, no. 1 (2009): 114, doi:10.1163/1477 28509X448993.
50. Colette Soler, "The Paradoxes of the Symptom in Psychoanalysis," in *The Cambridge Companion to Lacan*, ed. Jean-Michel Rabaté (Cambridge: Cambridge University Press, 2003), 92.
51. Eric Laurent, "The Purloined Letter and the Tao of the Psychoanalyst," in *The Later Lacan: An Introduction*, ed. Véronique Voruz and Bogdan Wolf (Albany: State University of New York Press, 2007), 30.
52. Diane Davis, *Inessential Solidarity: Rhetoric and Foreigner Relations*, Pitt Comp Literacy Culture (Pittsburgh, PA: University of Pittsburgh Press, 2010), 2.
53. Ibid., 68.
54. Soler, *Lacanian Affects: The Function of Affect in Lacan's Work*, 98.
55. Roberto Harari, *How James Joyce Made His Name: A Reading of the Final Lacan*, trans. Luke Thurston (New York: Other Press, 2002), 323.
56. Ibid., 121.
57. Prieto, "Writing the Subject's Knot," 174.
58. Jacques Lacan, "The Seminar of Jacques Lacan, Book XXIII: Joyce and the Sinthome, 1975–1976," trans. Cormac Gallagher (unpublished, n.d.), Session of May 11th, 1976, lacaninireland.com.
59. Edson Luiz André de Sousa, "An Ocean That Is Disarmed into Letters: Utopia and Psychoanalysis," in *The Epistemology of Utopia: Rhetoric, Theory and Imagination*, ed. Jorge Bastos da Silva (Cambridge: Cambridge Scholars Publishing, 2013), 160.
60. Prieto, "Writing the Subject's Knot," 174.
61. Harari, *How James Joyce Made His Name*, 324.
62. Prieto, "Writing the Subject's Knot," 175.
63. Harari, *How James Joyce Made His Name*, 122.

64. Dany Nobus, "Littorical Reading: Lacan, Derrida, and the Analytic Production of Chaff," *Journal for the Psychoanalysis of Culture & Society* 6, no. 2 (2001): 286.
65. Vitanza, *Negation, Subjectivity, and the History of Rhetoric*, 336.
66. Jacques Lacan, *The Ego in Freud's Theory and in the Technique of Psychoanalysis, 1954–1955 (Vol. Book II)*, ed. Jacques-Alain Miller, trans. Sylvana Tomaselli, The Seminar of Jacques Lacan 2 (New York: W. W. Norton & Company, 1991), 29.
67. Lacan, *On Feminine Sexuality*, 56.

Bibliography

Beckett, Samuel. 1991. *Three Novels: Molloy, Malone Dies, The Unnamable*. New York: Grove Press.

Blum, Virginia, and Anna Secor. 2011. "Psychotopologies: Closing the Circuit Between Psychic and Material Space." *Environment and Planning D: Society and Space* 29: 1030–1047. doi:10.1068/d11910.

Charraud, Nathalie. 2001. "Topology: The Möbius Strip between Torus and Cross-Cap." In *A Compendium of Lacanian Terms*, ed. Huguette Glowinski, Zita M. Marks, and Sara Murphy, Trans. Oliver Feltham and Dominique Hecq, 204–210. London: Free Association Books.

Chiesa, Lorenzo. 2006. "Count-As-One, Forming-Into-One, Unary Trait, S1." *Cosmos and History: The Journal of Natural and Social Philosophy* 2, nos. 1–2: 68–93.

Cresswell, Tim. 2013. *Geographic Thought: A Critical Introduction*. Malden: Wiley-Blackwell.

Davis, Diane. 2010. Inessential Solidarity: Rhetoric and Foreigner Relations, Pitt Comp Literacy Culture. Pittsburgh: University of Pittsburgh Press.

de Sousa, Edson Luiz André. 2013. "An Ocean That Is Disarmed into Letters: Utopia and Psychoanalysis." In *The Epistemology of Utopia: Rhetoric, Theory and Imagination*, ed. Jorge Bastos da Silva, 158–164. Cambridge: Cambridge Scholars Publishing.

Dow, Suzanne. 2014. "Lacan with Beckett." *Nottingham French Studies* 53, no. 1: 1–18.

Eidelsztein, Alfredo. 2009. *The Graph of Desire: Using the Work of Jacques Lacan*. Trans. Florencia F.C. Shanahan. London: Karnac Books.

Freud, Sigmund. 1989. *The Ego and the Id*. Trans. Joan Riviere. New York: W. W. Norton & Company.

———. 1994. *The Interpretation of Dreams*. Trans. A. A Brill. New York: Modern Library.

Friedman, N.A., and C.H. Séquin. 2007. "Keizo Ushio's Sculptures, Split Tori and Möbius Bands." *Journal of Mathematics and Art* 1, no. 1: 47–57. doi:10.1080/17513470701217217.

Gunn, Joshua. 2004. "Refitting Fantasy: Psychoanalysis, Subjectivity, and Talking to the Dead." *Quarterly Journal of Speech* 90, no. 1: 1–23. doi:10.1080/0033563042000206808.

Harari, Roberto. 2002. *How James Joyce Made His Name: A Reading of the Final Lacan*. Trans. Luke Thurston. New York: Other Press.

Lacan, Jacques. 1987. "Television." Trans. Denis Hollier, Rosalind Krauss, and Annette Michelson. *October* 40: 6–50.

———. 1991a. *Freud's Papers on Technique, 1953–1954*. Trans. John Forrester. The Seminar of Jacques Lacan, bk 1. New York: Norton.

———. 1991b. *The Ego in Freud's Theory and in the Technique of Psychoanalysis, 1954–1955 (Vol. Book II)*, ed. Jacques-Alain Miller and Trans. Sylvana Tomaselli. The Seminar of Jacques Lacan 2. New York: W. W. Norton & Company.

———. 1998. *The Four Fundamental Concepts of Psychoanalysis*. Trans. Alan Sheridan. Reiss. The Seminar of Jaques Lacan, Book 11. New York: Norton.

———. 1999. *On Feminine Sexuality: The Limits of Love and Knowledge*. Trans. Bruce Fink. The Seminar of Jacques Lacan, bk. 20. New York: Norton.

———. 2006. *Écrits: The First Complete Edition in English*. Trans. Bruce Fink. New York: W.W. Norton & Co.

———. 2009. "L'étourdit." Trans. Cormac Gallagher. *The Letter* 41: 31–80.

———. n.d. *The Seminar of Jacques Lacan, Book XXIII: Joyce and the Sinthome, 1975–1976*. Trans. Cormac Gallagher. unpublished, lacaninireland.com.

Laplanche, Jean, and Jean-Bertrand Pontalis. 2006. *The Language of Psychoanalysis*. Trans. Donald Nicholson-Smith. London: Karnac Books.

Laurent, Eric. 2007. "The Purloined Letter and the Tao of the Psychoanalyst." In *The Later Lacan: An Introduction*, ed. Véronique Voruz and Bogdan Wolf, 25–52. Albany: State University of New York Press.

Levy-Stokes, Carmela. 2001. "Jouissance." In *A Compendium of Lacanian Terms*, ed. Huguette Glowinski, Zita M. Marks, and Sara Murphy, 101–109. London: Free Association Books.

Lundberg, Christian. 2004. "The Royal Road Not Taken: Joshua Gunn's 'Refitting Fantasy: Psychoanalysis, Subjectivity and Talking to the Dead' and Lacan's Symbolic Order." *Quarterly Journal of Speech* 90, no. 4: 495–500.

———. 2012. *Lacan in Public: Psychoanalysis and the Science of Rhetoric*, Rhetoric, Culture, and Social Critique. Tuscaloosa: University of Alabama Press.

Marini, Marcelle. 1992. *Jacques Lacan: The French Context*. Trans. Anne Tomiche. New Brunswick: Rutgers University Press.

Miller, Jacques-Alain. 1994. "Extimité." In *Lacanian Theory of Discourse: Subject, Structure, and Society*, ed. Mark Bracher, Marshall W. Alcorn, Ronald J. Corthell, and Françoise Massardier-Kenney, 74–87. New York: New York University Press.

———. 2006. "Commentary on the Graphs." In *Écrits: The First Complete Edition in English*, ed. Jacques Lacan, and Trans. Bruce Fink. New York: W. W. Norton & Co.

Nobus, Dany. 2001. "Littorical Reading: Lacan, Derrida, and the Analytic Production of Chaff." *Journal for the Psychoanalysis of Culture & Society* 6, no. 2: 279–288.

———. 2003. "Lacan's Science of the Subject: Between Linguistics and Toplogy." In *The Cambridge Companion to Lacan*, ed. Jean-Michel Rabaté, 50–68. Cambridge: Cambridge University Press.

O'Donoghue, Diane. 2007. "Mapping the Unconscious: Freud's 'Topographic' Constructions." *Visual Resources* 23, nos. 1–2: 105–117.

Pile, Steve. 2014. "Beastly Minds: A Topological Twist in the Rethinking of the Human in Nonhuman Geographies Using Two of Freud's Case Studies, Emmy von N. and the Wolfman." *Transactions of the Institute of British Geographers* 39: 224–236. doi:10.1111/tran.12017.

Prieto, Graciela. 2011. "Writing the Subject's Knot." Transl. Kristina Valendinova. *Recherches En Psychanalyse* 12(December): 170–179. doi:10.3917/rep.012.0170.

Rickert, Thomas J. 2007. *Acts of Enjoyment: Rhetoric, Žižek, and the Return of the Subject*, Pittsburgh Series in Composition, Literacy, and Culture. Pittsburgh: University of Pittsburgh Press.

Soler, Colette. 2003. "The Paradoxes of the Symptom in Psychoanalysis." In *The Cambridge Companion to Lacan*, ed. Jean-Michel Rabaté, 86–101. Cambridge: Cambridge University Press.

———. 2016. *Lacanian Affects: The Function of Affect in Lacan's Work*. Trans. Bruce Fink. London: Routledge.

Tomšič, Samo. 2012. "Homology: Marx and Lacan." *S: Journal of the Jan van Eyck Circle for Lacanian Ideology Critique* 5: 98–113.

Vitanza, Victor J. 1997. *Negation, Subjectivity, and the History of Rhetoric*. Albany: State University of New York Press.

Weiss, Tzahi. 2009. "On the Matter of Language: The Creation of the World from Letters and Jacques Lacan's Perception of Letters as Real." *Journal of Jewish Thought & Philosophy* 17, no. 1: 101–115. doi:10.1163/147728509X448993.

CHAPTER 9

A Year of Deliberating Danger(ously): A Network Topology of the Loaded Climate Dice

Kenneth C. Walker

A foundational insight of classical rhetoric is that repeated patterns of discourse, or *topoi*, mark the formation and maintenance of communities. Yet, these communities and their associated *topoi* are sensitive to change when they enter opportune moments for rhetorical performance or *kairos*. *Kairoi* represent exigencies that place pressure on rhetorical performances to account for the shifting contexts within which actors try to persuade one another to change their attitudes, behaviors, and actions. As exigencies, *kairoi* provide a shorthand for describing a series of events that condition a given rhetorical performance and also circulate well beyond it. But for the rhetorical scholar, the specific problem is how to develop a method to analyze and/or intervene in the dynamics between *kairos* and situated performances that enact a rhetorical agency to form, maintain, and change communities. Topology, or a set of *topoi* that characterizes the discourse of a particular community, is one resolution to this methodological problem. Particularly when traced over time and across *kairoi*, topologies provide a concrete method for marking the rhetorical transformations within and across discourse communities.

K.C. Walker (✉)
Department of English, University of Texas at San Antonio,
San Antonio, TX, USA

© The Author(s) 2017

L. Walsh, C. Boyle (eds.), *Topologies as Techniques for a Post-Critical Rhetoric*, DOI 10.1007/978-3-319-51268-6_9

Rhetorical scholars have noted for some time now how these classical insights can be productively mapped onto Actor-Network Theory (ANT), which holds that the formation of political alliances is best described in terms of networks where agency is an effect of the material-relational system within which actors perform—including other actors, objects, technologies, institutions, ecological conditions, and other nonhuman entities.[1] ANT's notion of agency takes on various strong or weak claims to the agency of objects that cannot be fully rehearsed here.[2] For the purposes of this chapter, it will suffice to say that nonhumans have at least a mediating force on rhetorical performance, and one way to mark those mediations is to trace topologies across *kairoi* in order to discover how, when, and where topologies shift. Because they are concrete and identifiable discursive structures that maintain some stability while also undergoing change, topologies are powerful tools for the rhetorical analysis of networks, and particularly the process ANT describes as translation, or "[...] the work through which actors modify, displace, and translate their various and contradictory interests."[3]

This particular combination of rhetorical analysis and ANT can perhaps be best described as network topologies—an inductive method for mapping the rhetorical circulation processes over time and space for the analysis of material-discursive shifts in political alliances. Because network topologies are methodologically primed for analyzing the mediating effects of technologies on discourse, they are particularly well suited to study the politics of the Internet, a concern and methodology that shares some similarities to Rogers' network-topological approach.[4] In the case of climate risk communication on the Internet, communication scholars have shown how big-data approaches can provide a big picture of community formation by identifying repeated features of the structure and content of a discourse[5]; other scholars have used social media and content analysis to examine how skeptical blogs attempt to delegitimize and contest scientific expertise, and thus serve as alternative sites of public expertise for a climate-skeptical audience[6]; and others have examined user comments as a proxy for proto-deliberation and skeptical beliefs.[7] The network-topological analysis offered here is a practice situated between big-data approaches and individual comments on a single blog—a kind of medium-data approach appropriate for the study of deliberation. Through an analysis of a single case of climate risk communication online, the network topologies traced here represent "a slice through a data set"[8] that is specific enough to capture important aspects of rhetorical performance but distributed enough to gauge community formation and maintenance across nodes on a network (or sites on the Internet). This networked and case-based approach to climate risk performances provides a concrete rhetorical method for "zooming and hovering" when studying online deliberation.[9]

In order to demonstrate how a network-topological analysis can contribute to our understanding of how climate risk politics plays out on the Internet, this chapter traces the network of the popular risk metaphor of the loaded climate dice over one year across the English-written blogosphere. Once this network is mapped, I conduct a rhetorical analysis of uncertainty *topoi* based on the spheres model of uncertainty, which identifies that deliberation over risk is grounded in at least three integrated areas of activity: the technical, personal, and public.[10] Taken together these uncertainty *topoi* function as topologies for climate risk communication. These uncertainty topologies are important community markers because they provide insight into the motivations for interlocutors to enter into climate risk deliberations. Furthermore, this method can also help trace some of the networking capacities of the loaded climate dice metaphor as it circulates on the Internet.

The upshot of this network topology is an unfolding of the potentials and limitations of the loaded climate dice metaphor and agonistic proto-deliberations in online environments. While the metaphor is a well-crafted shorthand for political deliberation over climate risk, it has a limited capacity for community transformation because it instantiates a priori political values and removes important social judgments from public deliberation. Because the metaphor seeks to bypass rather than engage agonistic audiences, it cannot counter them directly. However, the proto-deliberations surrounding the uncertainty topologies do make explicit the core values of communities within climate risk deliberations. This surfacing of explicit value judgments can help confront manufactured scientific controversies, even if the dialogues are attenuated for factions of the American public who may have a conflicted sense of prudential action in regards to the science of climate risk.[11,12] Thus, while the actors surrounding the loaded climate dice metaphor seek to deliberate danger, its limitations mean that they may also deliberate dangerously by avoiding engagements through difference and reinscribing a wait-and-see approach.

Topologies of Uncertainty in a Year of Online Climate Risk Deliberations

Previous work has identified how a spheres approach to uncertainty arguments provides an important rhetorical method for understanding risk.[13] Briefly, that work identified how technical, personal, and public arguments function as a dynamic and interlocking set of *topoi* which when taken together form a topology (see Table 9.1 and Appendix A). I have adapted this model to ANT in order to show how topologies of uncertainty scaffold understandings of climate risk across different communities and different nodes in a network.[14] The take-away from that work is that topologies of uncertainty can usefully track how material-discursive performances legitimize interpretations of risk in public arguments that carry significant weight for actionable politics.

For climate politics in the US, 2012 was a signature year for online deliberation over climate risk. Since the economic collapse of 2008, climate risk was largely forgotten in the North American political imagination. Polls consistently showed US public ranked their concern over climate change near the bottom, and fewer people than before saw global warming as a threat.[15] With the defeat of cap and trade legislation in 2010, climate risk had become a political nonstarter. These issues came to a head during the 2012 presidential campaign, which came to be known for its climate silence during a year of extreme weather events. The last presidential debate was held on October 22 at Lynn University while Hurricane Sandy started brewing in the Caribbean Sea, later to hit the Eastern Seaboard on October 29, 2012, and launch climate risk back into the political scene. With little political acknowledgment of the record-breaking heat in the spring and extreme drought in the summer of that year,

Table 9.1 Uncertainty topoi and spheres of argument (see Walsh and Walker 2016)

Technical	Uncertainties addressed from disciplines that identify the limits of technical knowledge
Personal/epistemic	Uncertainties addressed on intimate scales that identify claims to commitment, confidence, concern, belief, and character
Public/political	Uncertainties address the collective and identify right action, shared or conflicting values, and potential harm

most of the deliberation over climate risk happened in the outer reaches of the mass media news cycle and the science blogosphere.

One of the few visible actors of a political climate rhetoric who did respond to these extreme weather events was the climatologist and climate advocate, Dr. James Hansen. As a stalwart of the climate debate, Hansen's performances in public are classic examples of what Lynda Walsh has called the prophetic ethos[16]—those performances of character that use access to specialized knowledge unavailable to common people to facilitate a political conversation about the values that public must recognize anew in order to make collectively informed decisions. Seasoned climate prophets like Dr. Hansen navigate the double ethical bind between technical expertise and political rhetoric with rhetorical tools to help them communicate uncertainties, contingencies, risks, and urgency in elegant fashion. Dr. Hansen's famous example, crafted over decades of climate campaigning, is the risk metaphor of the loaded climate dice. Risk metaphors conduct an extraordinary amount of rhetorical and political work. In this case, the loaded climate dice metaphor argues the more CO_2 that is pumped into the atmosphere, the more the climate dice are loaded toward extreme weather events. Thus, the metaphor frames CO_2 pollution as a high-stakes gamble where the odds of losing are increasingly predictable and the consequences increasingly severe.

In late 2011 Dr. James Hansen uploaded on his website a preprint of a report titled *Climate Variability and Climate Change: The New Climate Dice*. The report compared 30 years (1951–1980) of past surface temperature data on the entire globe to the latest 30 years (1981–2011), and it argued that statistically speaking, extreme weather events have become more common in the last 30 years. What climatologists call three-sigma (σ) events—extreme weather outliers that are more than three standard deviations warmer than previous climatology—now occur about 10 % more often. Hansen and his team concluded that the climate dice have progressively become more loaded and "thus there is no need to equivocate about the summer heat waves in Texas in 2011 and Moscow in 2010, which exceeded 3σ—it is nearly certain that they would not have occurred in the absence of global warming."[17] In this strategy Hansen bypasses controversies over climate models by using simple statistics to provide evidence for human-induced climate risk. And presumably simple statistics are more accessible and can be more appreciated by the public than tortuous climate models.

By all accounts 2012 was a remarkable year of extreme weather events across the globe, particularly in the US. The first half of 2012 witnessed extreme cold in Europe; extreme flooding across Brazil, China, and Australia; and monthly record-breaking spring heat in the US followed by a summer with unusually large fires, extended droughts, and more record-breaking heat across much of the country. In the fall of that year, countries across the Atlantic, including the Eastern Seaboard of the US, experienced hurricanes ranging from Jamaica to Maine, including Hurricane Sandy—the largest Atlantic Hurricane ever recorded and the second most costly, estimated at $ 75 billion (USD).[18] These extreme weather events broke the climate silence after the economic recession. During and after these multiple *kairoi* the US political rhetoric shifted, and it became possible to recognize and deliberate climate risk again.

Over the year the loaded climate dice metaphor circulated across the networked media landscape. This metaphor allowed climate communicators to acknowledge the increasing risks of extreme weather without falsely drawing a one-to-one, cause-and-effect relationship between climate change and any specific event. Yet, the ability of this circulation to foster risk deliberations and judgments remains an open question that network topologies of uncertainty are particularly well suited to address.

Proto-deliberations on Networked Media and the Agonism of Scientific Prudence

Networked media spaces such as science blogs provide an overwhelming amount of access to deliberation on climate risks. Yet access is attenuated by attention: though science blogs are open to all publics, they are often quasi-public spaces dominated by a few individuals; though science blogs are open deliberative forums, they function more like proto-deliberative spaces where arguments are tested before they are performed in deliberative spaces like courtrooms, policy hearings, and public debates.[19] In this way, while networked media amplifies and attenuates proto-deliberation, they still retain important continuities to fully deliberative spaces. In the case of climate science, the explosion of networked media in the public sphere has had roughly three effects: it has amplified the demand for sound-bite communication strategies; it has extended the long tail of science communication where almost any topic is easily accessible in extraordinary detail; and consequently, it has attenuated the impact of climate science in the competition for attention online.[20] So in online networked environments, information on climate risk is both made more available but also less important because of the competing demands in an attention economy. The study of how this metaphor allowed various public to enter into climate risk deliberations provides a case study in the work that risk metaphors play in gaining attention and shaping public perception of extreme weather and climate change in online environments.

As proto-deliberative spaces, science blogs are best understood not through the lens of agonistic deliberative rhetoric where strategic acts of communication are recognized, dissensus is valued, and yet interlocutors use discussion as an exploration of possible worlds and a renewal of identities.[21] Agonism is particularly characteristic of deliberations over climate risk because of its politicization as a manufactured scientific controversy by what some have termed a climate change denial industry.[22] While the full scale of the climate change denial industry cannot be rehearsed here, two components of it are important for studies of online climate risk deliberation. First, the denial machine links energy corporation funding to think tanks to the rare academic expert who is willing to talk openly about doubt, scientific uncertainty, and the lack of consensus about the science of climate change. The denial machine then amplifies the presence of these experts, which allows journalists to portray both sides as "balanced reporting" instead of a critical perspective on the issue.[23] Second, is astroturfing, or the manufacture of a grassroots organization that is often bankrolled by wealthy donors or think tanks. This plays out in many ways, but one is to get massive numbers of paid commentators—or trolls—to drown out online public discourse. So-called climate denier trolls arrive on the deliberative scene with an arsenal of strategies—they are often anonymous and go to extreme lengths to conceal their identity; they never engage the post in a substantial way but will often cherry-pick and attack one small, often technical, aspect of the author's argument and repeat. Other trolls use personal attacks or provocative comments to reduce the level of discussion. The effect is that trolls drive away engaged readers and make it difficult to tell who is genuinely skeptical about the science and who is mocking genuine skepticism to drag down what otherwise might be a good conversation. This has become such a problem for online forums that it gave rise to "resistance trolls"—science-savvy commenters who watch blogs and engage trolls in order to beef up the conversation with sound science. Some blog comments are nothing but paid trolls arguing with one another. The effect of astroturfing is a microcosm of the Internet more broadly: the arguments of trolls are amplified, yet the effects of the arguments are attenuated. Despite these problems, sometimes good conversations happen and research has shown that science blogs can facilitate real-world political action on science-related issues.[24,25]

What is at stake in an agonistic online environment is the battleground for the formation of arguments that gain attention, win adherents, and constitute emerging publics. It is possible then to view agonistic online climate risk discourse as a political resource for deliberation amidst difference. Rather than ending in consensus, the ends of agonistic deliberation are reflections and practices committed to "sustaining recognition and engagement even in a fractured, fractious, transient situation."[26] In this way, online trolling debates enact what is best and worst about controversies: at worst, controversies are a waste of time dealing with people who are (sincerely or not) in need of reliable information and continual correction; at best, controversies enrich democracy because they assemble desirable difference and disruptive social groups as political resources to reflect and renew commitments to core values, identities, and actions.[27]

In order to study these aspects of controversy, what follows is an analysis of the results from mapping the network of the loaded climate dice over a single year as a proxy for online climate risk deliberation.[28] Within this network I then analyze topologies of uncertainty using a dynamic view of the spheres model. The combination of tracing the network of the loaded climate dice and closely reading the uncertainty arguments therein is one way to study how this process of agonistic deliberative rhetoric plays out in online environments. Because topologies of uncertainties trace how actors map their ideological concerns onto online deliberations, they track the development of value judgments in personal decision making about risk. When these topologies are pulled across sites and over time, rhetoricians gain a better sense of how communities with various ideological stances respond to exigencies of extreme weather, and how they strategically engage in agonistic climate risk deliberations. They also usefully identify which sites are more or less contested in the battle for public deliberation in this controversy.[29]

Tracing the Networked Metaphor Across *Kairoi* and the Agonism of Climate Silence

Tracing the networked-risk metaphor across the web demonstrates that the circulation of the loaded climate dice metaphor mirrors the *kairoi* of extreme weather events during 2012. In January, when Dr. Hansen released an early draft of his study on the Internet, a few technical climate blogs picked it up and deliberated over its merits, largely in terms of technical uncertainties and scientific *ethos* (i.e., how credible is the study and the scientists behind it?). By April, climate denier blogs (*Watts Up With That*, and Roger Pielke Seniors' Blog) and an academic news site had picked up the metaphor.[30] But by July and August, during the extreme heat in the summer of 2012 in the US and elsewhere, sites across the web had picked up on the loaded climate dice. Similarly, the climate dice metaphor spiked again in late October and November in the wake of Hurricane Sandy. Thus, the *kairoi* of extreme weather events during 2012 prompted multiple deliberative events tracked through the loaded climate dice metaphor. Interestingly, online deliberations in the latter half of the year were characterized by a close association between technical and political uncertainty topologies, which was not true for the first half of the year. As one might expect, the extreme heat of late summer and Hurricane Sandy prompted more political deliberation over climate risk than the events of the spring and early summer.

Uncertainty topologies also usefully mark community formation across sites on the Internet, as well as sites where climate risk deliberation is most contested. In their big-data study, Elgesem, Steskal, and Diakopoulos identified that denier communities are most concerned with climate science and acceptor communities are mainly concerned with the politics of climate change rather than with climate science.[31] In my corpus that community phenomenon shows up in the topological data in denier blogs, climate blogs, and technical sites, the latter of which tend to discuss the issue more in terms of political uncertainty than technical uncertainty (see Table 9.2). Beyond science/climate blogs, news sites and opinion editorials had the most balanced of technical/public topologies of uncertainty, and also the highest number of comments. As I discuss later, high-profile opinion editorials are highly contestable sites for online climate risk deliberations with robust and highly agonistic comment sections. These sites are perhaps less about community formation than about battles for attention and relevance in attenuated online environments.

However, it is also noteworthy that no identifiably conservative news organization (Fox News, for example) in my year-long corpus picked up on the metaphor. While the corpus includes a broad range of political standpoints from sites like technical reviews, news reports, blogs, videos, editorials and more, news media with an identifiable conservative perspective are underrepresented. This finding corroborates what many commentators and viewers noticed during Hurricane Sandy: By and large, conservatively oriented media sites ignored any correlations between extreme weather events and climate change. Even the counter-examples

Table 9.2 Frequency of uncertainty *topoi* by forum

Total Corpus	Technical	Personal	Public
Posts (n = 1804)	3.2 % (n = 58)	0.8 % (n = 14)	3.3 % (n = 60)
Comment Section (n = 13,418)	1.3 % (n = 172)	0.7 % (n = 98)	0.8 % (n = 107)
Technical Sites	Technical	Personal	Public
Posts (n = 665)	2.0 % (n = 13)	0.6 % (n = 4)	3.0 % (n = 21)
Comment Section (n = 4049)	1.9 % (n = 75)	0.6 % (n = 24)	0.5 % (n = 19)
News Sites	Technical	Personal	Public
Posts (n = 585)	4.8 % (n = 28)	1.2 % (n = 7)	3.6 % (n = 21)
Comment Section (n = 1737)	2.0 % (n = 34)	2.0 % (n = 38)	1.2 % (n = 20)
Climate Blogs	Technical	Personal	Public
Posts (n = 244)	4.9 % (n = 12)	1.3 % (n = 2)	8.2 % (n = 20)
Comment Section (n = 438)	0.5 % (n = 2)	(n = 0)	0.7 % (n = 3)
Denier Blogs	Technical	Personal	Public
Posts (n = 166)	6 % (n = 10)	(n = 0)	1.2 % (n = 2)
Comment Section (n = 1214)	0.5 % (n = 6)	0.3 % (n = 4)	0.4 % (n = 5)
Opinion Editorials	Technical	Personal	Public
Posts (n = 144)	7.6 % (n = 11)	0.7 % (n = 2)	6.9 % (n = 10)
Comment Section (n = 6119)	0.9 % (n = 57)	0.3 % (n = 18)	0.9 % (n = 54)

Note: Uncertainty *topoi* were calculated at the sentence level. The frequency of uncertainty arguments represents the number of sentences with uncertainty arguments divided by the total number of sentences in the post or comment section.

Forums

Technical Sites/Blogs: Skeptical Science, NASA Technical Review, Climate, Etc., PNAS Review, Open Mind (Tamino), Harvard Business Review

News Sites/Blogs: Forbes, PBS News Hour, Northeastern News, NPR, Reader Supported News, The Conversation, Wall Street Journal

Climate/Environment Blogs: Climate Crocks, Climate Progress, Climate One, Scientific American, Mother Jones, National Geographic

Denier Blogs: Watt's Up With That?, GlobalWarming.org

Opinion Editorials: *New York Times*, *Washington Post;* CNN Opinion, *Huffington Post*

are instructive on this point: In the climate denial blog "Watts Up With That" (WUWT), the loaded climate dice are mentioned twice in just one post titled "Unloading James Hansen's Climate Dice." Both times it is used to deny anthropogenic causes, either through a character defamation of Dr. Hansen, or to associate warming trends with effects of urban heat islands.[32] Avoidance of the loaded climate dice metaphor in conservative mass media news sites is one way to negate deliberation; another is to dismiss the risk metaphor as a political rhetoric without any scientific legitimacy. Either way, climate silence is the result.

Uncertainty Topologies and Ideological Difference in Networked Deliberations

Examinations of rhetorical deliberations over the climate dice metaphor support previous findings that Internet forums are thoroughly hybridized; there is no deliberative space on the network—no genre, no blog, no website—where uncertainty arguments do not in some fashion intersect.[33] Throughout my corpus of 29 sites, uncertainty arguments constituted 7.5 % of the total discourse in the posts, and only 2.8 % of the discourse in the comment section. Deliberations around uncertainty were highest in climate blogs (15.7 % of posts) and opinion editorials (13.8 % of posts), and least in denier blogs (2.6 % of posts) and comment sections overall (2.8 % on average; see Table 9.2). Of the 7.5 % of uncertainties in the entire corpus of posts, about half were technical uncertainties (3.6 %) and half were public (3.3 %). These findings suggest that those writers who are crafting these discourses with the loaded climate dice metaphor are deliberately associating the limits of technical knowledge with calls for public action amidst potential harm. This is especially true in climate blogs. Just as the metaphor translates technical with political concerns through games of chance, the crafting of these posts reasons probabilistically through uncertainties in order to make progressive value judgments about the actions required *because* of the remaining climate uncertainties.

This topological hybridity plays out in a number of interesting ways. One can be found in technical blogs and websites that dominantly argued through technical appeals to the limits of knowledge, but also included arguments about the political need to act. In the wake of Hurricane Sandy in particular, some technical blogs such as *Skeptical Science* contained direct appeals to potential harm and collective action: "So far we have failed to achieve significant mitigation of greenhouse gas emissions, and as a result, *extreme weather events on steroids like Hurricane Sandy will cause more suffering than they would otherwise have, and we will have to adapt to a future in which these types of events occur more frequently.*"[34] This finding suggests that some networked technical forums or genres are more willing to draw associations between technical knowledge and political action that may be less common in traditional technical forums. In other words, some online forums that substantively engage the technical details of climate science (i.e., not popular climate blogs) still are willing to make their progressive political agendas explicit.

Conversely, news blogs and websites such as Forbes, PBS, NPR, The Conversation, and university news sites, showed a capacity to deliver more technical uncertainties than might be expected from similar studies with legacy media (see Table 9.2). Some of the most interesting quotes are those that hybridize arguments from all three *topoi* to demonstrate a mobility from technical warrants to public warrants that can at times frame scientific uncertainty as a cause for action: "*The little that we do understand, and our best experts, point to action required.* The temperature trend is currently 30% higher than Hansens [sic] 1981 projections which even then *caused concern* [...]. It just doesn't make any sense *to experiment with the only home we have.*"[35] These hybridized arguments are intriguing because while it is well established that in traditional news media we can expect arguments to move upstasis toward evaluation arguments, it is less well established that online news blogs and websites carry a larger capacity for technical discourse. The slight favoring of technical uncertainties in these networked public forums suggests this ability, but further research is needed to corroborate these findings, and network topologies are one way to conduct this work.

Still, it is widely acknowledged that legacy media embedded in networked environments remain some of the most important sites for online public deliberation, and in this corpus that is clearly the case with opinion editorials. Op-eds contained the most frequent deliberations over uncertainties (13.8 % in posts) and by far the highest number of comments. As represented in this corpus, they were written by identifiably progressive writers who reason through uncertainties to counsel a progressive scientific prudence. The comment sections sometimes followed the reasoning of these actors, for example, by linking conservative ideology with prudential action: "In the case of the environment that means taking a *conservative approach* to the risk of global warming. *Even if the climate change science is wrong the risk is so great that we need to be cautious and prudent.*"[36] A number of substantial comments thoroughly integrated these *topoi* into an argument for scientific prudence: "In any case, *we don't need absolute proof* that humans have caused the world to get hotter; we just need sufficient risk. Even a tiny *risk of a huge catastrophe justifies immediate preventive action.*"[37] Still, arguments from skeptics and deniers of all kinds are well represented. Hansen's opinion editorial in *The Washington Post* is a good example of how these forums are still powerful sites of deliberation in networked media: At the time of this research, his single post received over 27,000 comments, which in the opinions of some were "proof that the deluded and mercenary minions are being directed to 'hit' this page."[38] The larger number of comments on sites like these are worth further discussion in light of agonistic deliberative rhetoric and the politics of climate risk on the Internet.

The Agonistic Deliberative Rhetoric of Ongoing Engagement with Climate Deniers

The exchanges between skeptical/denier trolls and so-called resistance trolls are instructive for agonistic deliberative engagement and recognition across difference. In essence, these performances represent deliberative lessons from engaging those with whom you deeply disagree. For example, an exchange between Marc (an identifiable denier troll) and Alvin (the resistance troll) is indicative of how engaged climate risk deliberations tend to happen on the Internet. While comment mediators and other interlocutors suggest "do not feed the trolls," for good reasons, Alvin continually engages Marc in long technical discussions, for example, about how to gauge expertise and give trust in climate science. Alvin and Marc's continual engagement likely turns most readers away, but their agonistic deliberative exchange leads to some interesting revelations, as when Alvin writes to Marc: "I really understand that you don't like my presence here, which is why you sent an email to my director about this comment thread (he is okay with it by the way), but I will engage here because *at a personal level* I think it is really important that we do something now about AGW [Anthropogenic Global Warming]."[39] Here Alvin articulates the standard argumentative threads of progressive climate bloggers—reliable information leads to personal concern and therefore to policy action. On the other hand, Marc continues to attack the validity of Alvin's ideological position and his institutional position as a communications officer for the Australian Research Council.

Alvin's continual engagement does lead to a few moments of recognition and revelation between the two interlocutors: After a long discussion about what kind of climate denier Marc is (he does not deny anthropogenic climate change, but he thinks the impacts will not be that bad), and what kinds of policy measures he supports (improving transit networks, using nuclear instead of coal energy, improving solar) he articulates his primary point of agonism: "I do not support use of precious tax payer resources on the sort of research, (or is it propaganda, it is hard to tell the difference these days) being done by Ben Newell advertised here once again at the tax payers expense."[40] Eventually the comment thread leads to other denier and resistance trolls joining in the debate. The latter note that Marc's denier troll status is revealed by his disingenuous arguments, his personal attacks on others, and his unwillingness to identify himself. (Marc does not have a profile, a picture, or an active email while most others in the conversation do). One resistance troll in particular points this out continually and with increasing sarcasm. Needless to say, the upshot of these proto-deliberations is not agreement or consensus. But it is only through the ongoing engagement that interlocutors reveal the tactics of denier trolls like Marc, and articulate their personal motivations: "I am happy to put my name and images to my comments because *my current understanding of climate change gives me cause for concern for the future for my son*."[41]

When engaged effectively, the rhetorical joust between denier and resistance trolls does what Callon, Barthes, and Lascoumes acknowledge as the primary role of controversies: to use desirable difference and disruptive social groups as political resources to reflect and renew commitments to core knowledges, values, identities, and actions.[42] Agonistic deliberations among denier/resistance trolls makes those core values explicit so that they can be recognized, even if they are not shared. Thus, the ends of these deliberations are not consensus based, but commitments to sustaining recognition and engagement in fractious and transient situations.[43] This commitment to engage through difference is perhaps best summed up in this comment: "Yes, America, there is a 'wingnuttrollinstructor' sending forth his legions of vapid and often incompetent trolls. [...] And finally, [...] there are many on this thread of intelligence and good will who also know what the hell they're talking about. Thank you for fighting the good fight — see you at the next bout of insane wingnut trollery."[44] For these interlocutors, continual engagement with those with whom you deeply disagree is "the good fight" because they recognize that public discourse defines public subjectivities and communities. For all parties, the agonistic debate is tiring and attenuated, but it does clarify commitments to fundamental values. In this sense, agonistic trolling is productive for a political discourse characterized by manufactured scientific controversy because values can be made explicit through continual engagement. Thus, to counter the limitations of risk metaphors that bypass engagement, the issue moving forward will be how and when to best use these rhetorical strategies in offline deliberative forums.

Conclusion

This network topology of the climate dice metaphor through a single year of networked media puts forth a number of related arguments: first, topological methods incorporate argumentative contexts for identifying communities of practice in online environments. Here we saw how uncertainty topologies help identify accepter/denier communities and their hybridizations, and help identify denier/resistance trolls in comment threads. Second, uncertainty topologies are suggestive for uncovering forum/genre change and forum/genre hybridity on the Internet, and this utility was demonstrated in my analysis of technical blogs and news sites culminating in the identification of opinion editorials as a legacy forum/genre with a powerful networking capacity; third, agonistic deliberations between so-called denier and resistance trolls shows how a commitment to sustaining argumentative practices reinforces the contours of normative topology by making them recognizable in fractious rhetorical situations. These are the online rhetorical battlegrounds for testing arguments and rhetorical performances in spaces that have an attenuated capacity to form emerging publics.

These findings speak to the larger issue with risk metaphors like the loaded climate dice. On one hand, they are great shortcuts that bypass manufactured scientific controversies and immediately arrive at actionable politics. When backed by extreme weather events, these kinds of risk metaphors certainly can carry the potential to motivate change. On the other hand, the metaphor has some limitations: one may not know the dice are loaded until one has rolled the dice a few too many times. Hansen's proclamations that the dice *are* loaded are effective for those who share his progressive views on this issue, but they also remove important risk evaluations from public deliberation in favor of prudential judgments from the scientific community. In this sense, the risk metaphor still carries a technocratic impulse. Furthermore, climate risk is a wicked problem with no sense of linear cause and effect, which makes it nearly impossible to adjudicate blame. This motivates a gambler to keep gambling knowing full well he or she can escape the immediate threats from fire, flood, drought, and famine. Even with those events, the logic of the metaphor says the gambler likely won't experience the loss of extreme weather again. Those consequences are for "others." And even if one accepts that the climate dice are loaded, then everyone is to blame, which means no one is directly guilty, and the motivation for action is offset by a potentially dangerous wait-and-see approach. Therefore, it is the commitment to engage through difference that may represent our best hope for effective climate risk rhetoric.

Appendix A: Uncertainty Definitions by Sphere, Major Subcategories, and Sentence-Level Indicators of Uncertainty *Topoi*. Hybrids Contain More Than One Code

Sphere	*Definition*	*Sub-categories*	*Example from posts*	*Example from commentators*
Technical	Uncertainties addressed from disciplines that identify the limits of technical knowledge.	Ignorance Lack of Understanding	Although the technology of our world is shaped by probabilities, *we still do not understand statistical correlation intuitively.*	A different jet stream and modified blocking highs in turn change the tracks of depressions that often determine our daily weather. These developments *are not well understood* yet, but are amenable for study.
Personal/ epistemic	Uncertainties addressed on intimate scales that identify claims to commitment, confidence, concern, belief, and character.	Claims to confidence, concern, worry, belief, and trust	There is *medium confidence* that droughts will intensify in the 21st century in some seasons and areas, due to reduced precipitation and/or increased evapotranspiration.	There is *medium evidence and high agreement* that long-term trends in normalized losses have not been attributed to natural or anthropogenic climate change.
Public/ political	Uncertainties address the collective and identify right action, shared or conflicting values and potential harm.	Extreme and severe weather, Potential harm, Call to action	When we plotted the world's changing temperatures on a bell curve, *the extremes of unusually cool and, even more, the extremes of unusually hot* are being altered so they are becoming both *more common and more severe.*	What we are looking for now are arguments that will persuade vested interests that they must put aside those interests so we can all join together to provide the *maximum protection from the worst consequences a* nd an equitable sharing of the costs of addressing the less horrific ones.
Hybrids		"Lets get with it and *work on reasonable remedies and quit working* on proving that many of us *do not know how* to make reasonable statistical observations." (Pub/Tech) "In any case, *we don't need absolute proof* that humans have caused the world to get hotter; we just need sufficient risk. Even a tiny *risk of a huge catastrophe* justifies immediate preventive action" (Tech/Per/Pub). "While uncertainties remain in our understanding of climate science, we know enough to act now" (Tech/Pub).		

Notes

1. Bruno Latour, "The powers of association," In *Power, Action, and Belief: A New Sociology of Knowledge?*, ed. John Law, 264–280 (London: Routledge, 1986); Bruno Latour, "On recalling ANT," In, *Actor-Network Theory and After*, eds. John Law & John Hassard, 15–25. (Oxford: Blackwell, 1999); John Law, "Objects and spaces," *Theory, Culture, and Society*, 19, no. 5/6 (2002): 91–105.
2. For an overview, see Edwin Sayes, "Actor-Network Theory and Methodology: Just What Does it Mean to Say that Nonhumans Have Agency?" *Social Studies of Science*, 44, no. 1 (2013): 134–149.
3. Bruno Latour, *Pandora's Hope: Essays on the Reality of Science Studies.* (Cambridge, MA, Harvard University Press, 1999), 311.
4. Richard Rogers, "Mapping and the Politics of Web Space," *Theory, Culture, and Society,* 29, no. 4/5 (2002): 193–219.
5. Dag Elgesem, Lubos Steskal & Nicholas Diakopoulos, "Structure and Content of the Discourse on Climate Change in the Blogosphere: The Big Picture," *Environmental Communication*, 9, no. 2 (2015): 169–188.
6. Amelia Sharman, "Mapping the Climate Change Blogosphere," *Global Environmental Change*, 26, (2014): 159–170. doi:10.1016/j.gloenvcha.2014.03.003.
7. Luke Collins & Brigitte Nerlich, "Examining User Comments for Deliberative Democracy: A Corpus- driven Analysis of the Climate Change Debate Online," *Environmental Communication*, 9, no. 2 (2015): 189–207.
8. Lynda Walsh, "The Common Topoi of STEM Discourse," *Written Communication*, 27, no. 1 (2010): p. 125.
9. Damien Pfister, *Networked media, Networked Rhetorics: Attention and Deliberation in the Early Blogosphere,* (University Park, PA: Pennsylvania State University Press, 2014): 14.
10. Thomas Goodnight, "The Personal, Technical, and Public Spheres of Argument: A Speculative Inquiry into the Art of Public Deliberation," *Argumentation and Advocacy, 18,* (1982/2002): 214–227; Lynda Walsh & Kenneth C. Walker (2016) "Perspectives on Uncertainty for Technical Communication Scholars" *Technical Communication Quarterly*, 25, no. 2 (2016): 71–86.
11. Leah Ceccarelli, "Manufactured Scientific Controversy: Science, Rhetoric, and Public Debate," *Rhetoric and Public Affairs, 14,* no. 2 (2011): 195–228.
12. Robert Danisch, "Political Rhetoric in a World Risk Society," *Rhetoric Society Quarterly, 40,* no. 2, (2010): 172–192.
13. Ulrich Beck, *Risk society: Towards a new modernity.* (London: Sage, 1992); Lynda Walsh & Kenneth C. Walker (2016) "Perspectives on Uncertainty for Technical Communication Scholars" *Technical Communication Quarterly*, 25, no. 2 (2016): 71–86.

14. Kenneth Walker, "Mapping the Contours of Translation: Visualized Un/Certainties in the Ozone Hole Controversy," *Technical Communication Quarterly, 25,* no. 2 (2016): 104–120.
15. Frank Newport, "Americans' Global Warming Concerns Continue to Drop," *Gallup* (politics), March 11, 2010, http://www.gallup.com/poll/126560/Americans-Global-Warming-Concerns-Continue-Drop.aspx?g_source=&g_medium=&g_campaign=tiles.
16. Lynda Walsh, *Scientists as Prophets: A Rhetorical Genealogy,* (New York, Oxford University Press, 2013).
17. James Hansen, Makiko Sato, Reto Ruedy, "Public Perception of Climate Change and the New Climate Dice," *NASA Goddard Institute for Space Studies,* (2012), http://www.giss.nasa.gov/research/briefs/hansen_17/.
18. Hurricane/Post-Tropical Cyclone Sandy (Service Assessment), *United States National Oceanic and Atmospheric Administration's National Weather Service.* October 22–29, 2012 (Archived from the original on June 2, 2013). Retrieved August 14, 2016.
19. Lynda Walsh, "The Double-Edged Sword of Popularization: The Role of Science Communication Research in the Popsci.com Comment Shutoff," *Science Communication,* (2015): 1–12.
20. Paige Brown Jarreau, "Science Bloggers, and the Long-Tail of Science Writing," *Scilogs,* (June 6, 2014). Retrieved from http://www.scilogs.com/from_the_lab_bench/science-bloggers-and-the-long-tail-of-science-writing/.
21. Arabella Lyon, *Deliberative Acts: Democracy, Rhetoric, and Rights.* (University Park, Pennsylvania State University Press, 2013): 7.
22. Peter Jacques, Riley E. Dunlap, Mark Freeman, "The Organisation of Denial: Conservative Think Tanks and Environmental Scepticism," *Environmental Politics,* 17, no. 3 (2008): 349–385. doi:10.1080/09644010802055576.
23. While I have some reservations about the overt political language of "denier," I use it here to refer to actors who actively undermine climate science for ideological reasons, which is clearly the case in my analysis. A few varieties of denial include: deny the science, deny human cause, or deny impact is anything to worry about.
24. David Holmes, "Astroturfing the Climate Wars: Five Ways to Spot a Troll," *The Conversation*, October 13, 2013, http://theconversation.com/astroturfing-the-climate-wars-five-ways-to-spot-a-troll-19011.
25. Matthew Barton, "The Future of Rational-Critical Debate in Online Public Spheres." *Computers and Composition*, *22*, no. 2 (2005), 117–190; Justin Cheng, Cristian Danescu-Niculescu-Mizily, Jure Leskovec, "How Community Feedback Shapes User Behavior," *Association for the Advancement of Artificial Intelligence,* (2014), http://cs.stanford.edu/people/jure/pubs/disqus-icwsm14.pdf; Jeff Grabill, Stacey Pigg, "Messy Rhetoric: Identity Performance as Rhetorical Agency in Online Public Forums." *Rhetoric Society Quarterly*, 42, (2012): 99–119; Laura Gurak,

Smiljana Antonijevic, "Digital Rhetoric and Public Discourse." In *The Sage Handbook of Rhetorical Studies*, ed. A. A. Lunsford, K. H. Wilson & R. A. Eberly, 496–508 (Thousand Oaks, CA: Sage, 2009).

26. Lyon, *Deliberative* Acts, 23.
27. Michel Callon, Pierre Lascoumes, and Yannick Barthes, *Acting in an Uncertain World* (Cambridge: MIT Press, 2009): 33–34; Danisch, "Political Rhetoric."
28. I traced this network by searching for the phrase "loaded climate dice" and "climate dice" within the relevant year in multiple search engines. This produced 29 artifacts for analysis that I selected based on three criteria: the artifact had to at least mention loaded climate dice in the content of the post (not just the comments); the site had to show substantive deliberative engagement (either by engaging other posts, or through high number of substantive comments), and it had to have been posted within the relevant year. Once I had these artifacts, I organized them chronologically, read them broadly, and then conducted a rhetorical analysis with a primary focus on *kairos* and uncertainty *topoi*, while also keeping a close eye on the reception of the loaded climate dice metaphor. If they had them, I also coded the uncertainty arguments in the comment sections in order to get a sense of the deliberations happening there. In one case, because there were over 2700 comments, I only coded 10 % of the comments (n=270 comments).
29. Though the point cannot be fully developed here, this is why topologies have a particular lens on genre. As Miller (1984) and others have noted, genres represent typified responses to recurring situations. In the terms developed here we might say genres carry stable enough for now topologies. In climate blogs, topologies of uncertainty become typified by community because some are more willing to discuss varieties of uncertainties than others. Thus, topologies of uncertainty usefully mark community formation in this instance because they map onto ideological differences recurrent on given sites (denier blogs, science blogs, policy blogs, news blogs, etc.). While a full genre analysis of climate blogs is beyond the scope of this chapter, topologies are one useful way to track typified responses and also what may be unusual. This is one way topologies can contribute to big-data studies of online climate deliberations, and extend studies of single sites or comment threads. Perhaps of the most interest are sites where the greatest amount of topological hybridization happens because those sites it seems are highly contested.
30. Roger Pielke Sr.'s short post is represented in the WUWT post, so I did not include it in the corpus.
31. Elgesem, Steskal, Diakopoulos, "Structure and Content," 184.
32. Anthony Watts, "Unloading James Hansen's Climate Dice," *Watts Up With That?*, October 1, 2012, https://wattsupwiththat.com/2012/10/01/unloading-james-hansens-climate-dice/.

33. Categorizing these websites and blogs into forums is also a complicated process. While some sites are dominantly technical, they may be technical with a clear liberal policy bias (i.e., Skeptical Science) or a known tendency for climate skepticism (i.e., Climate). News websites are fairly easy to recognize, but then a site like The Conversation is run by academics "with journalistic flair." In this case, I chose to categorize it as a news site because the post was about current events, albeit also about probabilistic reasoning about climate risks. I separated out climate/environment blogs from climate denier blogs, mostly based on ideological differences, and opinion editorials were easy enough to categorize separately.
34. dana1981, November 1, 2012. "Hurricane Sandy and the Climate Connection," *Skeptical Science,* http://www.skepticalscience.com/hurricane-sandy-climate-connection.html. Italics mine.
35. Nick Kermode, April 16, 2012. Comment on Ben Newell, "If you want to roll the climate dice, you should know the odds," *The Conversation,* April 16, 2012, https://theconversation.com/if-you-want-to-roll-the-climate-dice-you-should-know-the-odds-6462#comment_32481. Italics mine.
36. Touko, July 22, 2012. Comment on Paul Krugman, "Loading the Climate Dice," *New York Times* (opinion), July 22, 2012, http://www.nytimes.com/2012/07/23/opinion/krugman-loading-the-climate-dice.html?_r=0. Italics mine.
37. Kerry Pechter, July 22, 2012. Comment on Paul Krugman, "Loading the Climate Dice," *New York Times* (opinion), July 23, 2012, http://www.nytimes.com/2012/07/23/opinion/krugman-loading-the-climate-dice.html?_r=0. Italics mine.
38. Walter Pearce, August 5, 2012. (6:13 p.m.), comment on James Hansen, "Climate change is here, and worse than we thought," *Washington Post* (opinion), August 3, 2012, https://www.washingtonpost.com/opinions/climate-change-is-here--and-worse-than-we-thought/2012/08/03/6ae604c2-dd90-11e1-8e43-4a3c4375504a_story.html?utm_term=.e0af67dd893c#comments.
39. Alvin Stone, April 16, 2012. Comment on Ben Newell, "If you want to roll the climate dice, you should know the odds," *The Conversation,* April 16, 2012, https://theconversation.com/if-you-want-to-roll-the-climate-dice-you-should-know-the-odds-6462#comment_32481. Italics mine.
40. Marc Hendrickx, April 16, 2012. Comment on Ben Newell, "If you want to roll the climate dice, you should know the odds," *The Conversation,* April 16, 2012, https://theconversation.com/if-you-want-to-roll-the-climate-dice-you-should-know-the-odds-6462#comment_32481. Italics mine.
41. Alvin Stone, April 16, 2012. Comment on Ben Newell, "If you want to roll the climate dice, you should know the odds," *The Conversation,* April 16, 2012, https://theconversation.com/if-you-want-to-roll-the-climate-dice-you-should-know-the-odds-6462#comment_32481. Italics mine.

42. Callon, Lascoumes, and Barthes, *Acting in an Uncertain World.*
43. Lyon, Deliberative Acts, 24.
44. Walter Pearce, August 5, 2012. (6:13 p.m.), comment on James Hansen, "Climate change is here, and worse than we thought," *Washington Post* (opinion), August 3, 2012, https://www.washingtonpost.com/opinions/climate-change-is-here--and-worse-than-we-thought/2012/08/03/6ae604c2-dd90-11e1-8e43-4a3c4375504a_story.html?utm_term=.e0af67dd893c#comments.

Bibliography

Barton, Matthew. 2005. "The Future of Rational-Critical Debate in Online Public Spheres." *Computers and Composition* 22, no. 2: 117–190.

Beck, Ulrich. 1992. *Risk Society: Towards a New Modernity*. London: Sage.

Callon, Michel, Pierre Lascoumes, and Yannick Barthes. 2009. *Acting in an Uncertain World*. Cambridge: MIT Press, 33–34.

Ceccarelli, Leah. 2011. "Manufactured Scientific Controversy: Science, Rhetoric, and Public Debate." *Rhetoric and Public Affairs* 14, no. 2: 195–228.

Cheng, Justin, Cristian Danescu-Niculescu-Mizily, Jure Leskovec. 2014. *How Community Feedback Shapes User Behavior*. Association for the Advancement of Artificial Intelligence. http://cs.stanford.edu/people/jure/pubs/disqus-icwsm14.pdf.

Collins, Luke, and Brigitte Nerlich. 2015. "Examining User Comments for Deliberative Democracy: A Corpus- Driven Analysis of the Climate Change Debate Online." *Environmental Communication* 9, no. 2: 189–207.

Danisch, Robert. 2010. "Political Rhetoric in a World Risk Society." *Rhetoric Society Quarterly* 40, no. 2: 172–192.

Elgesem, Dag, Lubos Steskal, and Nicholas Diakopoulos. 2015. "Structure and Content of the Discourse on Climate Change in the Blogosphere: The Big Picture." *Environmental Communication* 9, no. 2: 169–188.

Goodnight, Thomas. 1982/2002. "The Personal, Technical, and Public Spheres of Argument: A Speculative Inquiry into the Art of Public Deliberation." *Argumentation and Advocacy* 18: 214–227.

Grabill, Jeff, and Stacey Pigg. 2012. "Messy Rhetoric: Identity Performance as Rhetorical Agency in Online Public Forums." *Rhetoric Society Quarterly* 42: 99–119.

Gurak, Laura, and Smiljana Antonijevic. 2009. "Digital Rhetoric and Public Discourse." In *The Sage Handbook of Rhetorical Studies*, ed. A.A. Lunsford, K.H. Wilson, and R.A. Eberly, 496–508. Thousand Oaks: Sage.

Hansen, James, Makiko Sato, Reto Ruedy. 2012. "Public Perception of Climate Change and the New Climate Dice." NASA Goddard Institute for Space Studies. http://www.giss.nasa.gov/research/briefs/hansen_17/.

Holmes, David. 2013. "Astroturfing the Climate Wars: Five Ways to Spot a Troll." *The Conversation*, October 13. http://theconversation.com/astroturfing-the-climate-wars-five-ways-to-spot-a-troll-19011.

Jacques, Peter, Riley E. Dunlap, and Mark Freeman. 2008. "The Organisation of Denial: Conservative Think Tanks and Environmental Scepticism." *Environmental Politics* 17, no. 3: 349–385.

Jarreau, Paige Brown. "Science Bloggers, and the Long-Tail of Science Writing." *Scilogs*, June 6. http://www.scilogs.com/from_the_lab_bench/science-bloggers-and-the-long-tail-of-science-writing/.

Latour, Bruno. 1986. "The Powers of Association." In *Power, Action, and Belief: A New Sociology of Knowledge?* ed. John Law, 264–280. London: Routledge.

———. 1999a. *Pandora's Hope: Essays on the Reality of Science Studies*, 311. Cambridge, MA: Harvard University Press.

———. 1999b. "On Recalling ANT." In *Actor-Network Theory and After*, ed. John Law and John Hassard, 15–25. Oxford: Blackwell.

Law, John. 2002. "Objects and Spaces." *Theory, Culture, and Society* 19, no. 5/6: 91–105.

Lyon, Arabella. 2013. *Deliberative Acts: Democracy, Rhetoric, and Rights*, 7. University Park: Pennsylvania State University Press.

Newport, Frank. "Americans' Global Warming Concerns Continue to Drop." *Gallup* (politics), March 11. http://www.gallup.com/poll/126560/Americans-Global-Warming-Concerns-Continue-Drop.aspx?g_source=&g_medium=&g_campaign=tiles.

Pfister, Damien. 2014. *Networked Media, Networked Rhetorics: Attention and Deliberation in the Early Blogosphere*, 14. University Park: Pennsylvania State University Press.

Rogers, Richard. 2002. "Mapping and the Politics of Web Space." *Theory, Culture, and Society* 29, no. 4/5: 193–219.

Sayes, Edwin. 2013. "Actor-Network Theory and Methodology: Just What Does it Mean to Say that Nonhumans Have Agency?" *Social Studies of Science* 44, no. 1: 134–149.

Sharman, Amelia. 2014. "Mapping the Climate Change Blogosphere." *Global Environmental Change* 26: 159–170. doi:10.1016/j.gloenvcha.2014.03.003.

Walker, Kenneth. 2016. "Mapping the Contours of Translation: Visualized Un/Certainties in the Ozone Hole Controversy." *Technical Communication Quarterly* 25, no. 2: 104–120.

Walsh, Lynda. 2013. *Scientists as Prophets: A Rhetorical Genealogy*. New York: Oxford University Press.

———. 2015. "The Double-Edged Sword of Popularization: The Role of Science Communication Research in the Popsci.com Comment Shutoff." *Science Communication* 37, no. 5: 1–12.

Walsh, Lynda, and Kenneth Walker. 2016. "Perspectives on Uncertainty for Technical Communication Scholars." *Technical Communication Quarterly* 25, no. 2: 71–86.

CHAPTER 10

Getting Down in the Weeds to Get a God's-Eye View: The Synoptic Topology of Early American Ecology

Lynda Walsh and Lawrence J. Prelli

In this chapter, we take a step toward a topological theory of technical graphics. This theory leverages a foundational rhetorical principle—the *topos* or *locus*—to give an account of the structure of visual arguments within the disciplinary and political dynamics that gave rise to them—a topology. A topological approach can integrate verbal and visual modes of argumentation because *topoi* or *loci* operate in both modes. Furthermore, because they are always derived inductively from case analysis, a topological approach is flexible enough to apply to any rhetorical situation involving technical graphics. We demonstrate the analytical utility of topologies via a case study that discloses how the position of scientific observers shifted in early American ecology. Our analysis inductively situates early ecological graphics in their disciplinary and political dynamics. In the process, we discover a synoptic ecology that may provide an exemplar for others looking to reconstruct, if not also intervene in, similar cases.

L. Walsh (✉)
Department of English, University of Nevada, Reno, NV, USA

L.J. Prelli
Department of Communication, University of New Hampshire, Durham, NH, USA

© The Author(s) 2017

L. Walsh, C. Boyle (eds.), *Topologies as Techniques for a Post-Critical Rhetoric*, DOI 10.1007/978-3-319-51268-6_10

The Rhetoric of Technical Graphics

We are certainly not the first scholars to take a rhetorical approach to technical graphics. Critics from the field of technical communication have taken the rhetorical canons as a rough guide—particularly invention, arrangement, and style.[1] This work has yielded many useful analyses but not much theory-building, largely because its ultimate aim is often pedagogical and instrumentalist—that is, advising STEM researchers and students how to produce "more effective" technical graphics.[2] Also, a small cadre of rhetoricians has worked more closely with the canonical vocabulary of classical rhetoric.[3]

The few scholars from technical communication and rhetoric who have gone on to build theory have largely resorted to frameworks outside rhetoric—primarily forms of social semiotics borrowed from science studies, but also principles from gestalt and cognitive psychology.[4] In principle, there is nothing wrong with extradisciplinary approaches to a rhetoric of technical graphics—as long as the resulting theory can account for the structure and reveal the political effects of technical graphics in situ. In practice, extradisciplinary attempts to assemble a coherent visual rhetoric of technical graphics have gotten mired in two related problems.

First, assembling a rhetoric from nonrhetorical materials often requires so much bricolage that the theory becomes unwieldy. Secondly, and more importantly, in the welter of extradisciplinary concepts, rhetoricians often lose sight of *kairos*. *Kairos*, as explained in the glossary to this volume, can either draw our attention to noticeable ruptures in business-as-usual or remind us that business-as-usual is really a continuous flow of ruptures smoothed over by habit and craft.[5] Whichever facet of *kairos* rhetoricians choose to emphasize—and we emphasize the second—it is this central tension between craft and chance that animates our discipline and distinguishes it from the other disciplines that work on technical graphics. Thus, these problems with top-heaviness and kairotic insensitivity may explain why rhetoricians of technical graphics have yet to rally behind a theory for them.

As a potential and partial solution to these problems, we propose topologies. They offer several attractive benefits to scholars of technical graphics. First, just as verbal rhetorics operate on verbal grammars, topologies can operate on any extant grammar of images, such as gestalt principles or Kress and Van Leeuwen's popular semiotic grammar.[6] Second, topologies are simple, assembled from exactly two foundational rhetorical concepts—*kairos* and *topos*. Third, topologies are flexible, accommodating unproblematically the mixing of visual, verbal, and other modalities of argumentation. Fourth, topologies are inductive, requiring the critic to work with and against *kairos* while reconstructing persistent communal attitudes or habits.

The introduction to this volume has already explained how topologies bridge from text to politics, so we will not rehearse those arguments. However, we should review the arguments justifying the application of *topoi*—which have traditionally been assumed to be verbal—to visual arguments. *Topoi* find their roots in the pre-Socratic tradition as terms describing places or regions. In fact, Rubinelli argues that in the fourth century BCE the most common use of the term was actually to denote a military position.[7] So, Aristotle inherited the *topos* as visual, spatial, and dynamic—"stance" more than "place." But when he adopted it as his organizing metaphor for rhetorical invention—finding in the chaotic throes of a given *kairos* the available stances most likely to yield success—he lent the concept a verbal, rational coloration that, over subsequent centuries of interpretation, became dominant. In short, *topoi* became understood as logical formulae when in fact they had deep roots in nonverbal and nonrational processes. Put back on their original historical footing, *topoi* are best understood as ambivalent "machines"[8] of rhetorical invention that may take verbal or visual expression (or aural, gestural, tactile, kinesthetic, etc.).

This understanding of *topoi* prepares us to interpret technical graphics topologically, by which we mean first to interpret graphics using the *kairoi* of their presentation to force to the surface the *topoi* from which the graphics were invented, and then to observe how those *topoi* interacted to constitute political and disciplinary relationships. In the following sections we take this topological approach to a historical sequence of related *kairoi*. We track the changing locus of the ecological observer as visualized in tables, maps, and photographs during the period when ecology was emerging as a scientific field in the United States.

The metaphors of "places" or "stances" are particularly apt to our case as early ecology was entirely absorbed with places and how to position the scientific observer to best understandingly "see" them. We examine the *topoi*—both verbal and visual—of these technical graphics as they unfold from the vantage of viewers rather than of the authors.[9] Via these *topoi*, viewers are situated in specific observational stances that determine how facts may be constituted. The view from any such stance is partial and, thus, necessarily persuasive and political. So, when we disclose the verbal and visual choices encoded by specific *topoi* of viewing, we reveal how ways of looking are constituted and enabled, thus revealing important political dynamics in the emergence of ecology as a science.

Shifts in Topologies of Observation in Early Ecology

Our case study examines how *topoi* enabled and structured distinct ways of constituting observed facts in Roscoe Pound and Frederick E. Clements's collaborative work, which has been widely credited with ushering in important methodological and mathematical developments in the emergent science of ecology.[10] We will detail how some of these developments can be traced to *topoi* that reconstituted the facts of ecology and how to observe them.

Our story begins with Roscoe Pound's enthusiastic review of Oscar Drude's *Deutschlands Pflanzengeographie* (1896). Pound proclaimed that Drude's "geographico-botanical survey of a large country, on a large scale, will be a great inspiration" for those hard at work on "biological and botanical surveys."[11] Pound and his associate Clements were then at work on a large-scale project of their own, a dissertation they would later publish as *The Phytogeography of Nebraska* (1900).[12] They lauded Drude's exacting thoroughness and his focus on which plant species were dominant in a particular landscape.[13] Indeed, Pound argued that plant biologists needed to shift attention from identifying and classifying individual plant species to what we could call an integrative if not holistic approach to plant assemblages. That shift was only recently underway, initiated by efforts to "collate" and "group" facts systematically so that "biological and physiological facts" could readily be applied to them. "Comparison" of these "biological groups" was enabled by sophisticated uses of statistics. That is when the "work acquired importance."[14] Pound explained:

> To understand the vegetation of a region one must ascertain not only what are its physical, meteorological and geological features, but much more what sorts of plants control its water, meadow, plain, or forest vegetation. Directed toward the latter ends, statistics have a very different meaning. Such work is the aim of the new geographical botany.[15]

Pound and Clements offered a method that presumed eyewitness observation of plant assemblages in the field would all-too-often become overly impressionistic. "In determining the abundance of species, appearances are extremely deceptive," Pound and Clements wrote. "Long continued observation in the field stamps a picture on one's mind, and it seems a simple matter to pick out the several species and to classify them in the several grades of abundance with reasonable accuracy," they continued. But the matter was not so simple. Even though field experience impressed the mind with "mental pictures" that seemingly made "the reference of the commoner secondary species of prairie formations to their proper grades an easy task," Pound and Clements concluded that those classifications often proved "largely erroneous."[16]

In their argument, Pound and Clements invited their readers to perceive a methodological issue in terms of the *topos* of *appearance versus reality*. All too often, methods of observation were imprecise and impressionistic; they did not always furnish reliable facts. If the problem was too much reliance on appearances, then the remedy must be disclosure of empirical reality. That is what Pound and Clements proposed to accomplish with their new method of observation.

Transcendence Through Quantification

To transcend the limitations of a human scientific observer situated in the landscape, Pound and Clements literally got down in the weeds. They counted the individuals of secondary species found in five-square-meter sample plots located both within plant formations and in transition zones between them. (This is called the "quadrat" method.) They presented their results in tables, which provide actual counts of secondary species found in plots located about 400 yards apart within the prairie grass formation (*Sporobolus-Koeleria-Panicum*) (see Fig. 10.1).

The tables invite viewers to see listed species from the general *topos* of *degree*, or *more or less*. Species are arranged top to bottom from the highest to the lowest numbers on each of the two tables. The viewer is further encouraged to process the two orders of abundance comparatively by seeing them according to the general *topos* of *similarity and difference*. The result: the top nine species are the same for the two tables, and the counts for each species on the two tables are comparable.

At first glance, this seems obvious, but Pound and Clements's method ushered in an important transition in the vantage of observing and in the facts observed. The act of counting individual plants at the strategically located plot sites converts the observed facts from sensually encountered phenomena into abstract numbers via the disciplinary *topos* of *quantification*. Once quantified, those observations become "data" fit for statistical manipulation.

Amorpha canescens	387
Aster multiflorus	223
Antennaria campestris (16 patches)	209
Solidago rupestris	101
Helianthus rigidus	97
Kuhnistera candida	43
Kuhnistera purpurea	31
Brauneria pallida	24
Solidago rigida	19
Kuhnia glutinosa	8
Comandra umbellata	7
Rosa arkansana	2

(2)

Amorpha canescens	368
Aster multiflorus	395
Antennaria campestris (6 patches)	154
Solidago rupestris	105
Helianthus rigidus	63
Kuhnistera candida	26
Kuhnistera purpurea	22
Brauneria pallida	19
Solidago rigida	16
Rosa arkansana	8
Solidago rigidiuscula	5
Gerardia purpurea	3
Laciniaria scariosa	2
Erigeron ramosus	1

Fig. 10.1 Quantifying the observable. Originally unnumbered tables, Pound and Clements, "A Method of Determining the Abundance of Secondary Species," 20–21

For example, Pound and Clements calculate averages. They point out that the top counts shown on the two tables are confirmed by averaging counts from "a large number of plots." The data from those plots show that *Amorpha canescens* has an average of 309 individuals and that *Aster multiflorus* has an average of 275 individuals. Similarly, *Antennaria campestris* finishes third, with 145 individuals in "12 patches."[17] It is important to note that those numbers do not merely describe phenomena, as would the more common practice of measuring the dimensions of plant features.

Nor are those numbers adduced to illustrate an argument. Rather, the statistical averages *constitute the argument*; they mathematically demonstrate the relative abundance of the top three secondary species of the prairie grass formation (*Sporobolus-Koeleria-Panicum*).[18] These statistical manipulations thus yield new scientific facts.

Pound and Clements argued that their method could increase observational precision by creating a numerical basis for distinguishing levels of abundance. Drude had distinguished abundance levels according to the descending categories of "social," "gregarious," "copious," "sparse," and "scarce" (using the general *topos* of *degree* or *more or less*).[19] Classification according to these levels was based upon the trained botanist's sense observations. Tobey aptly calls this approach "qualitative quantification."[20] In naming the ordinals of his scale, Drude leaves traces of the *topos* of the sociological discipline, *society*. Viewing plants as social actors in different "formations" or "associations" was common at this time, and Drude's scale locates the scientific observer amid the throng, so to speak.

Pound and Clements instead chose to classify "degrees" of abundance using averages of secondary species counted at their plots. To illustrate, consider the "copious" category that, like the others, stands for "a quantitative idea as well as the manner of association of the individuals."[21] Copious in the first degree or $copious^1$ contains more than 200 individuals, $copious^2$ 150–200 individuals, $copious^3$ 100–150 individuals, and so on. Pound and Clements thus combined the general *topoi* of *division* and *degree* to replace Drude's qualitative ordinal scale with a more precise, quantitative interval scale.

This new scale performs the disciplinary *topoi* of *precision* and *accuracy*. These are the special strengths of Pound and Clements's "plot and counting" methodology. For example, the new distribution establishes the observation that species classified as subcopious "are often very striking components of prairie formations."[22] Subcopious species can *appear* very abundant, but use of the plot method discloses *realities* that show otherwise. The statistical method corrects the scientific observer's sensory biases.

The five-square-meter-plot method not only provided an accurate way to disclose relative abundance of secondary species in a vegetation area, it yielded data that could be used to stake out the boundaries of plant formations. Pound and Clements presented tables showing data for decline in number and relative abundance of secondary species peculiar to a for-

Amorpha canescens	291
Aster multiflorus	238
Kuhnistera candida (?)	23
Solidago rupestris	21
Brauneria pallida	17
Helianthus rigidus	12
Kuhnia glutinosa	5

Fig. 10.2 Detecting unseen boundaries. Originally, unnumbered table, Pound and Clements, "A Method of Determining the Abundance of Secondary Species," 21

mation in adjacent transition areas, through which "we have been able to make more accurate limitations of the regions and particularly the transition areas than we had thought possible" (Fig. 10.2).[23] Here, again, the general *topoi* of *degree* and *similarity and difference* structure readings of the tables. Consider, for example, the raw counts from a plot sited in the transition area between the prairie grass formation and the sand hill region.

Based on implied comparison with the tables in Fig. 10.1, the two scientists employ the general *topos* of *degree* when they tell the reader: "The marked decrease in the number of secondary species and in the abundance of each is characteristic of this transition area."[24] This and two other tables—constructed with the help of *quantity*, *similarity and difference*, and *degree topoi*—yield an abstract view of regional plant life through simple "enumeration."

This quantitative transcendence dislocates the scientific observer from the landscape by shifting the vantage of observation from the situated to the abstract, from the sensual to the formal. This transcendent "view from nowhere," to borrow Donna Haraway's terminology,[25] is the key move in terms of the politics of the ecological imaginary during this epoch. Quantitative transcendence helps relocate the observer to a stance from which he purportedly can see and comprehend the entire landscape at a glance: the synoptic move.

The Synoptic Move

The final step to a truly scientific ecological observer is accomplished by the action of spatial *topoi*. Pound and Clements attempt verbally to evoke the distinctive contours of vegetation formations that the abstract numbers can only at best imply:

> The constant diminution in the number and abundance of secondary species as one passes from the prairie grass formation of the prairie region to the buffalo grass formation of the transition area and of the sand hill region is well illustrated by these figures. The difference between the prairie grass formation in its ordinary situation and in the transition area, and between the buffalo grass formation of the "range" and the same formation in the transition area is better shown by figures obtained from such enumerations than in any other way.[26]

Spatial *topoi* are also used to depict bounded features visually, as we see when Pound and Clements set out to map the prairie province and its constituent features.

Pound and Clements wrote an article that extrapolated the boundaries of major formations comprising the North American prairie province based in part on data from their work, *Phytogeography of Nebraska*. As they put it: "the data have been extended ... to cover the entire prairie province, while in the work referred to only the territory embraced in the political limits of Nebraska is considered."[27] With "province" the encompassing term, Pound and Clements characterized its constituent "regions," thereby evoking general *whole-to-part* or *part-to-whole topoi*. They argued that Drude's previous work had mischaracterized the prairie province; due to "inaccurate data," Drude's account of the Missouri prairie was "erroneous as to boundary and as to characterization."[28] Pound and Clements set out to offer more accurate, bounded characterizations.

Pound and Clements articulate the prairie province's features in relation to four regions: "I. Wooded bluff and meadowland region; II. Prairie region; III. Sand hill region; IV. Foothill region."[29] Readers discover quickly that region I—the "Wooded bluff and meadowland region"—is not part of the prairie province at all, as Drude had erroneously suggested.[30] The other three regions, Pound and Clements maintained, were "well-differentiated" features of the central plains (Fig. 10.3).[31]

This figure offers a map of the prairie province's three constituent regions that evokes a top-down, god's-eye angle of observation. Note that the province is bounded by the large broken lines. Small broken lines separate the contiguous regions within the province. These regions are inscribed over a political palimpsest of states, provinces, and sovereign

Fig. 10.3 Visualizing boundaries: Explanation of "Plate XXI. Sketch map of the 'Prairie Province:' I. the prairie region; II. the sand hill region; III. the foothill region." From Pound and Clements, "Vegetation Regions of the Prairie Province," 394

Indian territories. One is invited to read the map's features in terms of *parts and wholes*: states are parts of the United States and provinces are parts of Canada; at the same time, the three vegetative regions are parts of the whole that is the prairie province. The natural boundaries, legitimated by science, cut across the boundaries of political administration.

Thus, science visually stakes a claim to the Continent over and against sovereign governments based upon its superior vantage for observing real, natural boundaries.

Pound and Clements's three regions are depicted visually and described verbally, with each modality reinforcing the other. First, coordinates are given: "The prairies proper (including those of Iowa and Illinois)" [region I], for example, are seen "extending from the forest to the 98th meridian." That region, in turn, is "easily distinguished from the sand hills" [region II]. The sand hills region has a "general western boundary between 102–103° W," which, in turn, makes it "very distinct from the foothills" [region III].[32] Regional boundaries are further detailed as follows:

> To the south, in the Indian territory, the prairies proper are crowded out by the sandy plains, and the prairie region disappears, leaving regions [II] and [III]. On the Saskatchewan plains, the forests close in on the east, and the sandy plains drop out, resulting in a similar reduction, although here it is region [II] which disappears. ... As a consequence of the stronger development of region [I] to the northward and of region [II] to the southward, the division of the prairie province gives it a peculiar dovetailed appearance. The prairie region tapers gradually to the south, finally disappearing in the Indian territory. The sand hill region likewise narrows toward the north, but much more rapidly, vanishing in the Dakotas, and attaining only occasional expression to the northward. The submontane region [III] skirts the base of the Rocky mountains from Athabasca to New Mexico.[33]

The god's-eye view is thoroughly established. Terms of spatial orientation ("south," "northward," "southward"), shape ("dovetailed appearance," "tapers," "narrows"), and place ("Indian territory," "Dakotas," "New Mexico," "Athabasca," "Rocky mountains") presume a viewer positioned to comprehend regions of thousands of square miles. And this viewer is able for the first time to see global movements. The prairies are "crowded out," the forests "close in," the sandy plains "drop out." We are told about the "stronger development" of one region compared to others. Visually, features appear as static items on a map, but verbal depictions invite us to view them as dynamic entities.

Verbally Inducing Unseen Bodies

Clements would praise Henry Chandler Cowles for his contributions to "developmental" ecology. His work "championed" the study of large plant formations undergoing development from earlier to advanced stages.[34] Clements himself was a leading champion of this approach when he advocated the view that plant formations are organisms developing through successive lifecycle phases, a view that would later be targeted in Henry Allan Gleason's subsequently influential criticisms of treating plant associations as though they are distinct "natural units" classifiable by type.[35] Interestingly, Cowles used two-dimensional photographs to infer not only the dimension of perspectival depth, but also especially motion. The viewer of the photographs is invited to watch plant communities move. Consider the "Pond-Swamp-Prairie" zonation process in Fig. 10.4.

The distinct "zones" are delineated verbally and visually by species, with the visual indicator of relative "lightness" used to mark them out. We are told in the original caption to see "[s]edges ... encroaching on the lake, while shrubs ... are encroaching on the sedges." The latter point is further elaborated in the text. "*Fig. 19* shows that a coniferous vegetation, now represented

Fig. 10.4 Making the static move. Originally, Fig. 19, from Cowles, "The Physiographic Ecology of Chicago and Vicinity," 148

by but two or three small trees at the centers of the islands, is to follow the Cassandra," Cowles wrote. "Such an advance of conifers on Cassandra is shown in the background at the right."[36] We are invited to see these static photographic elements as though captured in movement, with the implication that sedges will fill in the pond, shrubs will replace the sedges, and conifers will replace the shrubs.

How to show plant formations undergoing processes like zonation or succession sets a special challenge for visual depiction. Here we find Cowles's attempt to meet that challenge by verbally prompting viewers to "see" spatial features as undergoing movement. Vocabulary such as "encroach" and "advance" evokes associations of dynamism about static features on a photograph. This verbally induced sense of *motion* works on visual contrasts between "lighter" and "darker" areas, which manifest the general locus of *similarity and difference.*

A topology of motion was essential to the formulation of a dynamic, developmental ecology. And since motion and life have been equated since Aristotle, this topology also laid the groundwork for an organismic ecology that Clements would come to champion. That ecology depended on constructing plant communities as integrated, developing organisms with the implication that scientists are the only observers with the proper stance, locus, or perspective for "seeing" them move and undergo developmental changes

Conclusions

We have shown how early American ecological observations shift from direct, situated encounters with observed phenomena at different levels of precision, to statistical abstractions stripped of initial naturalistic contexts, to technically mediated forms of observation involving synoptic spaces on maps and static features on photographs made dynamic through verbally induced associations of movement. These shifts were generated with visual and verbal *topoi* that created a new topology of observation. This topology bore several interrelated implications for the politics of early American ecology and continues to bear implications for scholars studying related cases.

First, consider the varied topical traces disclosed via our inductive analysis of Pound and Clements's publications over a series of *kairoi* that together instigated a shift from naturalistic to abstract topologies of observation. There is the pivotal locus, appearance versus reality. There are general *topoi*: degree (more or less), similarity and difference, part and whole, and motion. We encountered disciplinary values that legitimate claims: quantification, precision, and accuracy. Finally, there are specific terminologies that impute specific meanings to the phenomena observed. Plants are conceived in sociological terms: the social and gregarious, individuals, associations, and formations. They also are arranged and grouped as though they are units of political geography: provinces, regions, and districts. This assemblage of *topoi* increasingly invited readers/viewers into a synoptic vantage from which to attend to, conceive, and observe plant life.

The synoptic shift in Pound and Clements's work created far-reaching effects. Its influence on the management of public lands was nearly immediate. Clements wrote: "The first application of quantitative methods to the problems of grazing came in 1905 with the organization of the national forests and the necessity of finding a basis for grazing allotments, which led to the use of quadrats in connection with reconnaissance and especially on the grazing reserves that were soon established."[37] He argued further that ecological science was at the foundation of public land policy, including national forest and public range management.[38] The quadrat in some variation became part of "the customary procedure in all quantitative studies of vegetation."[39] And due to "the growing appreciation of quantitative values," ecological methods were expanded in use in the fields of forestry, agriculture, grazing, and other areas.[40] We discern in Clements's comments the genesis of the natural resource expert, whom John S. Dryzek characterized as an "administrative rationalist" who believes that environmental issues and problems should be left for the experts to solve.[41]

The synoptic shift from sensory observation to the quantification of "fact" is a powerful source of the expert's rhetorical power. Pound and Clements's proposed new method promised the thrust needed to launch the observer beyond his limitations in the field and ecology into the stratosphere of the true sciences. But ironically, this dislocation could only be effected through the intensive application of *topoi*: chiefly appearance/reality, quantification, similarity/difference, degree, and part-to-whole. In other words, Haraway's scientific "view from nowhere" is upon closer inspection a specifically rhetorical achievement; it is generated by a topology so intricate that it makes locality, situation, and subjectivity appear to disappear.

Similar subject-dislocating topologies have been documented in disciplines also seeking scientific status during this historical epoch, such as economics, statistics, and cartography.[42] In these cases as well as in ours, synoptic topologies serve to stake science's claim to our shared habitat, certifying scientific observers to govern while disqualifying traditional authorities. Indeed, we suspect that it has contributed to formation of the ecologist's ethos as something akin to a progressive era scientist-planner. Consider the rhetorical power of that ethos in relation to natural resource issues. The ecologist can observe facts that are often unavailable to other observers. This special ability calls into question the experiential ethos of others who engage directly with the resource in question. We suspect that this is a frequent source of the well-documented tensions among technical administrative experts and ranchers, foresters, and fishers that persist to the present day.

Interestingly, however, Pound and Clements's consequential move "up" to synopticism required a preliminary move "down" to quantify the individual. This quantification reduced plants' individual bodies to numbers that could be combined to generate a larger ecological body to observe and govern. In other words, in their rhetorical attempts to transcend Drude's work, Pound and Clements put ecology on a path toward biopolitics.[43]

The generative typology we have limned from our case study thus yields a stance in discourse that we call synoptic ecology. The scientific observer is positioned as the only subject able to comprehend ecological bodies and thus govern them. This has implications for studies well beyond our case study area. Perhaps the most serious of these is the implication that global problems can only be fixed by technocratic solutions (e.g., climate engineering). Further, synoptic topologies, with their associated discourses of expertise, continue to powerfully constrain not only ecological discourses, but also discourses of health, medicine, economics, and education. Rhetoricians working on emergent discourses in these fields may wish to watch for the "move down" that presages a "move up" of quantitative transcendence and the invention of a synoptic topology.

For example, rhetoricians working on economic globalization would be well served to attune themselves to synoptic topologies because they also provide a framework for inventing alternative topologies. In fact, it was by disclosing the topology of "bioprospecting" in Amazonian rainforests that indigenous activists were recently able to imagine and deploy a strategic counter-rhetoric of "biopiracy" that successfully defended their territories against exploitation by multinational pharmaceutical companies.[44] In these ways and others, analyzing *topoi* and topologies can help rhetoricians reveal how the discourses they generate determine who and what are seen in and who and what are erased from our shared political habitat. Such disclosures are prelude to imagining alternative topologies for those discourses.

Notes

1. Paul Dombrowski, "Ernst Haeckel's Controversial Visual Rhetoric," *Technical Communication Quarterly* 12, no. 3 (2003); Miles Kimball, "London through Rose-Colored Graphics: Visual Rhetoric and Information Graphic Design in Charles Booth's Maps of London Poverty," *Journal of Technical Writing and Communication* 36, no. 4 (2006); Kathryn M. Northcut, "The Making of Knowledge in Science: Case Studies of Paleontology Illustration," *Dissertation Abstracts International, Section A: The Humanities and Social Sciences* 65, no. 3 (2004); Carol Reeves, "Scientific Visuals, Language, and the Commercialization of a Scientific Idea: The Strange Case of the Prion," *Technical Communication Quarterly* 20, no. 3 (2011); Derek G. Ross, "Dam Visuals: The Changing Visual Argument for the Glen Canyon Dam," *Journal of Technical Writing and Communication* 38, no. 1 (2008).
2. Eva Brumberger, "Making the Strange Familiar: A Pedagogical Exploration of Visual Thinking," *Journal of Business and Technical Communication* 21, no. 4 (2007); J. Dinolfo, B. Heifferon, and L.A. Temesvari, "Seeing Cells: Teaching the Visual/Verbal Rhetoric of Biology," *Journal of Technical Writing and Communication* 37, no. 4 (2007); S. Dragga and D. Voss, "Cruel Pies: The Inhumanity of Technical Illustrations," *Technical Communication* 48, no. 3 (2001); Charles Kostelnick and Michael Hassett, *Shaping Information: The Rhetoric of Visual Conventions* (Carbondale: Southern Illinois University Press, 2003).
3. For rhetorical analyses using classical figures, see Jeremiah Dyehouse, "'A Textbook Case Revisited': Visual Rhetoric and Series Patterning in the American Museum of Natural History's Horse Evolution Displays," *Technical Communication Quarterly* 20, no. 3 (2011).For *endoxa*, see Cara Finnegan, "The Naturalistic Enthymeme and Visual Argument: Photographic Representation in the 'Skull Controversy,'" *Argumentation & Advocacy* 37, no. Winter (2001); Lynda Walsh, "'Tricks,' Hockey Sticks, and the Myth of Natural Inscription: How the Visual Rhetoric of Climategate Conflated Climate with Character," in *Image Politics of Climate Change: Visualizations, Imaginations, Documentations*, ed. Thomas Nocke and Birgit Schneider (Bielefeld: Springer, 2014). For structures related to general topoi see Ben F. Barton and Marthalee S. Barton, "Modes of Power in Technical and Professional Visuals," *Journal of Business and Technical Communication* 7, no. 1 (1993); Dennis Dake, "Aesthetics Theory," in *Handbook of Visual Communication: Theory, Methods, and Media*, ed. Ken Smith, et al. (Mahwah, NJ: Lawrence Erlbaum Associates, 2005); Mary Rosner, "Theories of Visual Rhetoric: Looking at the Human Genome." *Journal of Technical Writing & Communication* 31 (2001).

4. Lee E. Brasseur, *Visualizing Technical Information: A Cultural Critique* (Amityville, NY: Baywood Publishing, 2003); Heather Graves, "The Rhetoric of (Interdisciplinary) Science: Visuals and the Construction of Facts in Nanotechnology," *Poroi* 10, no. 2 (2014); Alan G. Gross and Joseph E. Harmon, *Science from Sight to Insight: How Scientists Illustrate Meaning* (Chicago, IL: University of Chicago Press, 2013); Chad Wickman, "Observing Inscriptions at Work: Visualization and Text Production in Experimental Physics Research," *Technical Communication Quarterly* 22, no. 2 (2012).
5. See for the contextual and continual facets of *kairos* Debra Hawhee, "Kairotic Encounters," in *Perspectives on Rhetorical Invention*, ed. J. Atwill and J.M. Lauer (Knoxville, TN: University of Tennessee Press, 2002); Thomas Rickert, "Location in the Wild: On Locating *Kairos* in Space-Time," in *The Locations of Composition*, ed. C.J. Keller and C.R. Weisser (Albany, NY: State University of New York Press, 2007). See also Bruno Latour's notion of "hiatus" and visual representation as a technique for bridging it in "How to Be Iconophilic in Art, Science, and Religion," in *Picturing Science, Producing Art*, ed. Caroline Jones and Peter Galison (New York: Routledge, 1998), 431, and his *An Inquiry into Modes of Existence* (Cambridge, MA: Harvard University Press, 2013), 91.
6. Gunther Kress and Theo Van Leeuwen, *Reading Images: The Grammar of Visual Design* (New York: Routledge, 1996, 2006).
7. Sara Rubinelli, *Ars Topica: The Classical Technique of Constructing Arguments from Aristotle to Cicero* (Dordrecht, NL: Springer, 2009), 14.
8. Ibid.
9. A brief note on verbal/visual interaction in our case: there has been an active debate over whether or not images can "argue" in the absence of words. For example, see David Fleming, "Can Pictures Be Arguments?," *Argumentation &and Advocacy* 33, no. 1 (1996); Alan G. Gross, "Toward a Theory of Verbal-Visual Interaction: The Case of Lavoisier," *Rhetoric Society Quarterly* 39, no. 2 (2009).This debate is mooted by technical graphics, which virtually never appear unaccompanied by verbal or numerical text. Thus, we assume that viewers construct topologies of technical graphics from visual and verbal traces among other sources. We acknowledge the modality of the topical traces we encounter in our case (e.g., the word "contrast" versus a visual contrast between dark and light areas). However, we remain unconvinced that maintaining a philosophical or psychological distinction between word and image yields results in rhetorical studies of technical graphics from a topical or topological vantage—so we do not dwell on the distinction.
10. Frank Benjamin Golley, *A History of the Ecosystem Concept in Ecology: More Than the Sum of the Parts* (New Haven, CT: Yale University Press, 1993), 19; Robert P. McIntosh, *The Background of Ecology: Concept and Theory*

(Cambridge, UK: Cambridge University Press, 1985), 131–132; Ronald C. Tobey, *Saving the Prairies: The Life Cycle of the Founding School of American Plant Ecology, 1895–1955* (Berkeley, CA: University of California Press, 1981), 48–75.
11. Roscoe Pound, "The Plant-Geography of Germany," *The American Naturalist* 30, no. 354 (June 1896): 468.
12. Roscoe Pound and Frederic E. Clements, *Phytogeography of Nebraska: I. General Survey*, 2nd ed. (Lincoln, NE: Seminar., 1900).
13. Pound and Clements, *Phytogeography of Nebraska: I. General Survey*, 2nd ed., 4.
14. Pound, "The Plant-Geography of Germany," 465.
15. Ibid., 466.
16. Roscoe Pound and Frederic E. Clements, "A Method of Determining the Abundance of Secondary Species," *Minnesota Botanical Studies* 2 (1898): 19.
17. Ibid., 21.
18. According to Tobey, a paper from Charles E. Bessey, Pound and Clements's teacher, signaled an important shift in the use of numbers when he offered tables containing relative percentages of plant types. We find Pound and Clements doing similar kind of work. Tobey's point that Bessey used statistics *as arguments* applies as well to Pound and Clements's use of statistics. See Charles E. Bessey, "Phylogeny and Taxonomy of the Angiosperms," *Botanical* Gazette 24 no. 3 (Sep. 1897); Tobey, *Saving the Prairies: The Life Cycle of the Founding School of American Plant Ecology, 1895–1955*, 55–56.
19. *Saving the Prairies, Tobey*, 52.
20. Ibid., 48, 52.
21. Pound and Clements, "A Method of Determining the Abundance of Secondary Species," 23.
22. Ibid., 24.
23. Ibid.
24. Ibid., 21.
25. Donna Haraway, "Situated Knowledges: The Science Question in Feminism and the Privilege of Partial Perspective," *Feminist studies* 14, no. 3 (1988): 589.
26. Pound and Clements, "A Method of Determining the Abundance of Secondary Species," 22.
27. Roscoe Pound and Frederic E. Clements, "The Vegetation Regions of the Prairie Province," *Botanical Gazette* 26, no. 6 (1898): 387.
28. Ibid., 383.
29. Ibid., 387.
30. Ibid.
31. Roman numerals I–IV used in the article's text do not correspond with the three mapped "regions" designated as I–III on the prairie province map

provided as Plate XXI. After reviewing the maps appended near the end of *Phytogeography of Nebraska*, we suspect that the numerical sequence applied to a political map of Nebraska's four regions, and the two scientists somehow confused the two. The feature designated as numeral I in the article does not appear on the prairie province map of the three regions for the simple reason that it is not one of the province's constituent regions. Regardless, we have corrected the text so that the Roman numerals correspond with the regions on the map provided: region I designates the prairie region, region II refers to the sand hill region, and region III denotes the foothill region.

32. Pound and Clements, "The Vegetation Regions of the Prairie Province," 387.
33. Ibid., 387–388.
34. Frederic E. Clements, "Plant Succession. Pub. 242," (Washington: Carnegie Institute of Washington, 1916), 28.
35. Henry A Gleason, "The Individualistic Concept of the Plant Association," *Bulletin of the Torrey Botanical Club* 53, no. 1 (Jan. 1926).
36. Henry Chandler Cowles, "The Physiographic Ecology of Chicago and Vicinity: A Study in the Origin, Development, and Classification of Plant Societies (Conclusion)," *Botanical Gazette* 31, no. 3 (Mar. 1901), 149.
37. Frederic E. Clements, "Experimental Ecology in the Public Service," *Ecology* 16, no. 3 (Jul. 1935), 351.
38. Ibid., 349–354.
39. Ibid., 347.
40. Ibid., 349.
41. John S. Dryzek, *The Politics of the Earth: Environmental Discourses*, 3rd ed. (Oxford: Oxford University Press, 2013), 75–98.
42. Cf. the "god-trick" from Donna J. Haraway, *Modest_Witness@Second_Millennium. Femaleman(©)_Meets_Oncomouse™: Feminism and Technoscience* (New York: Routledge, 1997), 135–36; "mechanical objectivity" from Peter Galison, "Judgment against Objectivity," in *Picturing Science, Producing Art*, ed. Caroline Jones and Peter Galison (New York: Routledge, 1998), 355; and "expertise" from Theodore Porter, *Trust in Numbers: The Pursuit of Objectivity in Science and in Life* (Princeton, NJ: Princeton University Press, 1995), 7–8.
43. For a framing of the discussion of biopolitics in technical disciplines, see Michel Foucault, *The Birth of Biopolitics: Lectures at the Collège De France, 1978–1979* (Springer, 2008); Donna Haraway, "The Biopolitics of Postmodern Bodies: Determinations of Self in Immune System Discourse," *Feminist theory and the body: A reader* 1, no. 1 (1999). For examples of rhetorical analyses of biopolitics in technical disciplines see for economics Catherine Chaput, "Rhetorical Circulation in Late Capitalism: Neoliberalism

and the Overdetermination of Affective Energy," *Philosophy and Rhetoric* 43, no. 1 (2010); and for medicine Lisa Keränen, "Biopolitics, Contagion, and Digital Health Production: Pathways for the Rhetoric of Health and Medicine," *Communication Quarterly* 63, no. 5 (2015).

44. See chapter three of Leah Ceccarelli, *On the Frontier of Science: An American Rhetoric of Exploration and Exploitation* (MSU Press, 2013).

Bibliography

Barton, Ben F., and Marthalee S. Barton. 1993. "Modes of Power in Technical and Professional Visuals." *Journal of Business and Technical Communication* 7, no. 1: 138–162.

Bessey, Charles E. 1897. "Phylogeny and Taxonomy of the Angiosperms." *Botanical Gazette* 24, no. 3: 145–178.

Brasseur, Lee E. 2003. *Visualizing Technical Information: A Cultural Critique*. Amityville: Baywood Publishing.

Brumberger, Eva. 2007. "Making the Strange Familiar: A Pedagogical Exploration of Visual Thinking." *Journal of Business and Technical Communication* 21, no. 4: 376–401.

Ceccarelli, Leah. 2013. *On the Frontier of Science: An American Rhetoric of Exploration and Exploitation*. East Lansing: MSU Press.

Chaput, Catherine. 2010. "Rhetorical Circulation in Late Capitalism: Neoliberalism and the Overdetermination of Affective Energy." *Philosophy and Rhetoric* 43, no. 1: 1–25.

Clements, Frederic E. 1916. *Plant Succession. Pub. 242*. Washington: Carnegie Institute of Washington.

———. 1935. "Experimental Ecology in the Public Service." *Ecology* 16, no. 3: 342–363.

Cowles, Henry Chandler. 1901. "The Physiographic Ecology of Chicago and Vicinity: A Study in the Origin, Development, and Classification of Plant Societies (Conclusion)." *Botanical Gazette* 31, no. 3: 145–182.

Dake, Dennis. 2005. "Aesthetics Theory." In *Handbook of Visual Communication: Theory, Methods, and Media*, ed. Ken Smith, Sandra Moriarty, Gretchen Barbatsis, and Keith Kenney, 3–22. Mahwah: Lawrence Erlbaum Associates.

Dinolfo, J., B. Heifferon, and L.A. Temesvari. 2007. "Seeing Cells: Teaching the Visual/Verbal Rhetoric of Biology." *Journal of Technical Writing and Communication* 37, no. 4: 395–417.

Dombrowski, Paul. 2003. "Ernst Haeckel's Controversial Visual Rhetoric." *Technical Communication Quarterly* 12, no. 3: 303–320.

Dragga, S., and D. Voss. 2001. "Cruel Pies: The Inhumanity of Technical Illustrations." *Technical Communication* 48, no. 3: 265–274.

Dryzek, John S. 2013. *The Politics of the Earth: Environmental Discourses*. 3rd ed. Oxford: Oxford University Press.

Dyehouse, Jeremiah. 2011. "'A Textbook Case Revisited': Visual Rhetoric and Series Patterning in the American Museum of Natural History's Horse Evolution Displays." *Technical Communication Quarterly* 20, no. 3: 327–346.

Finnegan, Cara. 2001. "The Naturalistic Enthymeme and Visual Argument: Photographic Representation in the 'Skull Controversy.'" *Argumentation & Advocacy* 37, no. 3: 133–149.

Fleming, David. 1996. "Can Pictures Be Arguments?" *Argumentation & Advocacy* 33, no. 1: 11–22.

Foucault, Michel. 2008. *The Birth of Biopolitics: Lectures at the Collège De France, 1978–1979*. Basingstoke: Springer.

Galison, Peter. 1998. "Judgment Against Objectivity." In *Picturing Science, Producing Art*, ed. Caroline Jones and Peter Galison, 327–359. New York: Routledge.

Gleason, Henry A. 1926. "The Individualistic Concept of the Plant Association." *Bulletin of the Torrey Botanical Club* 53, no. 1: 7–26.

Golley, Frank Benjamin. 1993. *A History of the Ecosystem Concept in Ecology: More Than the Sum of the Parts*. New Haven: Yale University Press.

Graves, Heather. 2014. "The Rhetoric of (Interdisciplinary) Science: Visuals and the Construction of Facts in Nanotechnology." *Poroi* 10, no. 2: Article 12.

Gross, Alan G. 2009. "Toward a Theory of Verbal-Visual Interaction: The Case of Lavoisier." *Rhetoric Society Quarterly* 39, no. 2: 147–169.

Gross, Alan G., and Joseph E. Harmon. 2013. *Science from Sight to Insight: How Scientists Illustrate Meaning*. Chicago: University of Chicago Press.

Haraway, Donna. 1988. "Situated Knowledges: The Science Question in Feminism and the Privilege of Partial Perspective." *Feminist Studies* 14, no. 3: 575–599.

———. 1997. *Modest_Witness@Second_Millennium.Femaleman(©)_Meets_Oncomouse™: Feminism and Technoscience*. New York: Routledge.

———. 1999. "The Biopolitics of Postmodern Bodies: Determinations of Self in Immune System Discourse." *Feminist Theory and the Body: A Reader* 1, no. 1: 203.

Hawhee, Debra. 2002. "Kairotic Encounters." In *Perspectives on Rhetorical Invention*, ed. J. Atwill and J.M. Lauer, 16–35. Knoxville: University of Tennessee Press.

Keränen, Lisa. 2015. "Biopolitics, Contagion, and Digital Health Production: Pathways for the Rhetoric of Health and Medicine." *Communication Quarterly* 63, no. 5: 504–509.

Kimball, Miles. 2006. "London Through Rose-Colored Graphics: Visual Rhetoric and Information Graphic Design in Charles Booth's Maps of London Poverty." *Journal of Technical Writing and Communication* 36, no. 4: 353–381.

Kostelnick, Charles, and Michael Hassett. 2003. *Shaping Information: The Rhetoric of Visual Conventions*. Carbondale: Southern Illinois University Press.

Kress, Gunther, and Theo Van Leeuwen. 1996 [2006]. *Reading Images: The Grammar of Visual Design*. New York: Routledge.

Latour, Bruno. 1998. "How to Be Iconophilic in Art, Science, and Religion." In *Picturing Science, Producing Art*, ed. Caroline Jones and Peter Galison, 418–440. New York: Routledge.

———. 2013. *An Inquiry into Modes of Existence*. Cambridge, MA: Harvard University Press.

McIntosh, Robert P. 1985. *The Background of Ecology: Concept and Theory*. Cambridge, UK: Cambridge University Press.

Northcut, Kathryn M. 2004. "The Making of Knowledge in Science: Case Studies of Paleontology Illustration." *Dissertation Abstracts International, Section A: The Humanities and Social Sciences* 65, no. 3: 917.

Porter, Theodore. 1995. *Trust in Numbers: The Pursuit of Objectivity in Science and in Life*. Princeton: Princeton University Press.

Pound, Roscoe. 1896. "The Plant-Geography of Germany." *The American Naturalist* 30, no. 354: 465–468.

———. 1898. "The Vegetation Regions of the Prairie Province." *Botanical Gazette* 25, no. 6: 381–394.

———. 1900. *Phytogeography of Nebraska: I. General Survey*. 2nd ed. Lincoln: Seminar.

Pound, Roscoe, and Frederic E. Clements. 1898. "A Method of Determining the Abundance of Secondary Species." *Minnesota Botanical Studies* 2: 19–24.

Reeves, Carol. 2011. "Scientific Visuals, Language, and the Commercialization of a Scientific Idea: The Strange Case of the Prion." *Technical Communication Quarterly* 20, no. 3: 239–273.

Rickert, Thomas. 2007. "Location in the Wild: On Locating *Kairos* in Space-Time." In *The Locations of Composition*, ed. C.J. Keller and C.R. Weisser, 61–90. Albany: State University of New York Press.

Rosner, Mary. 2001. "Theories of Visual Rhetoric: Looking at the Human Genome." *Journal of Technical Writing & Communication* 31: 391–413.

Ross, Derek G. 2008. "Dam Visuals: The Changing Visual Argument for the Glen Canyon Dam." *Journal of Technical Writing and Communication* 38, no. 1: 75–94.

Rubinelli, Sara. 2009. *Ars Topica: The Classical Technique of Constructing Arguments from Aristotle to Cicero*. Dordrecht: Springer.

Tobey, Ronald C. 1981. *Saving the Prairies: The Life Cycle of the Founding School of American Plant Ecology, 1895–1955*. Berkeley: University of California Press.

Walsh, Lynda. 2014. "'Tricks,' Hockey Sticks, and the Myth of Natural Inscription: How the Visual Rhetoric of Climategate Conflated Climate with Character." In *Image Politics of Climate Change: Visualizations, Imaginations, Documentations*, ed. Thomas Nocke and Birgit Schneider, 55–81. Bielefeld: Springer.

Wickman, Chad. 2013. "Observing Inscriptions at Work: Visualization and Text Production in Experimental Physics Research." *Technical Communication Quarterly* 22, no. 2: 150–171.

CHAPTER 11

Enthymematic Elasticity in the Biomedical Backstage

Christa Teston

> We can articulate rich physical phenomena using notions like the wash of ripples along the banks of a river, the accumulation of leaves in the eddies trapped in the crook of a tree trunk fallen into the water, or more symbolic entities like the destinations of lanterns set out to float on the current, or the origins of a river and all its tributaries.
>
> —Xin Wei Sha, "Topology and Morphogenesis"

Carving at the marrow of medical uncertainty requires marshaling many kinds and forms of evidence. But the technoscientific practices integral to manufacturing medical evidence are constantly evolving, as are the environmental, political, and sociocultural conditions that shape human health in the first place. Definitions of, and contexts for discovering medical evidence undergo continual change. Some characterize such change as "progress." Progress in the breast cancer community, for example, has resulted in more than 60 modifications to treatment standards in the last 20 years alone. Amid biomedicine's technoscientific progress, however, are ongoing threats to corporeal stability and security. Alarmingly high numbers of newly diagnosed, life-threatening diseases occur every year, especially among those disproportionately affected by "failing social and

C. Teston (✉)
Department of English, Ohio State University, Columbus, OH, USA

© The Author(s) 2017

L. Walsh, C. Boyle (eds.), *Topologies as Techniques for a Post-Critical Rhetoric*, DOI 10.1007/978-3-319-51268-6_11

economic networks of support."[1] Given biomedicine's constant flux and in light of precarious corporealities,[2] how do medical professionals navigate medical uncertainty amid constantly changing phenomena?

After having spent the better part of a decade studying decision-making and evidential construction in cancer care,[3,4,5] I can confidently assert that each case study yields one general conclusion: medical professionals consistently use enthymematic argumentation to navigate medical uncertainty. That is, cancer care professionals work together to mobilize properties of one domain about which they are reasonably certain so as to help elucidate another domain about which they are less certain. I understand enthymematic argumentation as a topological tactic, or a method for opening up inquiry (and thereby, prognostic possibility) amid the corporeal chaos caused by cancer. To illustrate how enthymemes operate in the biomedical backstage,[6] in this chapter, I describe cancer care professionals' rhetorical tactics by employing as my analytic frame the inventive power of topological structures.

For the purposes of this project, I draw on Walsh and Boyle's definition of topologies as providing communities of practice with "reflexes for dealing with uncertainty" (13). As the editors of this volume indicate, many topological theories have their roots in mathematics. It may seem counterintuitive to mobilize a mathematical construct as a way to demonstrate the dynamic nature of enthymematic argumentation amidst pervasive flux and corporeal contingencies. However, consider Sha's use of the ephemerality of smoke (and the co-construction involved in bearing witness to it) to describe topologies as "flow" in dynamic systems:

> Stand a group of people in a room; ask someone to light and smoke a cigarette. Ask each person to raise a hand upon smelling the smoke. This seems like a reasonable way to empirically define where the smoke is. But notice several features about this experiment. The extent of the smoke changes with time. The extent is determined physiologically, situationally, phenomenally: different people have different sensibilities and each person may be more or less sensitive to smoke according to how much s/he thinks about the smoke. In fact, just asking people to smell for smoke primes their sensitivities. Therefore the smoke's extent is an amalgam of the physical particles in motion, the people's physiologies, and the phenomenological expectation set by the asking.[7]

So while *topoi* are, as the editors to this volume argue in the introduction, concrete and traceable, they are also, as illustrated in Sha's above illustration, ephemeral and contingent. Ikoniadou reminds readers that topology is "the science of proximity."[8] In other words, topological theory assumes that things are never truly fixed "but, rather, will follow an incessant process of *(de)*formation, where the boundaries of concrete and abstract, inside and outside, become flexible (vibratory) and thus continuously redrawn."[9] Topological analyses can also provide rhetoricians with a way to mine arguments for the value-laden conditions that politicize them in particular ways.[10]

Mobilizing topologies as a critical analytic unearths how allopathic medical professionals in the US (in particular, those who practice cancer care) navigate medical uncertainty by leveraging the suasive power of enthymematic argumentation.[11] Drawing on Dyck,[12] I situate *topoi* as enthymematic premises that are, according to Scott,[13] architectonically productive. Identifying and tracing *topoi* requires attunement to the flow of material-discursive conditions that structure argumentation. While layered and complex, such attunements have tremendous payoff. By locating and mapping the evolution of enthymematic premises, topological analyses reveal the commitments to which certain communities of practice adhere.

In what follows, I begin with contemporary definitions for enthymematic argumentation so that readers might more readily see the relationship between enthymemes and *topoi*. What emerge in that discussion are distinct characterizations of enthymemes as logical, probabilistic arguments versus characterizations of enthymemes as flexible and capable of responding to local conditions. Taking to be true the latter characterization—that is, enthymematic argumentation is more about plausibility than probability—I provide an example of how enthymematic argumentation unfolds in the biomedical backstage. Through a topological analysis of cancer care professionals' deliberation about a patient's particularly complicated case, I demonstrate that evidence-based arguments are neither dogmatic nor doctrinaire.

In fact, results from my topological analysis of deliberations in the biomedical backstage suggests that when cancer care professionals mobilize medical evidences toward some argumentative end, they must straddle several often times competing, co-constructed evidential premises—a practice Susan Leigh Star calls "tacking back-and-forth" between "ill-structured and well-structured aspects."[14] Borrowing from Star, therefore, I argue that cancer care professionals perform medical expertise by tacking back and forth between statistical and standardized medical evidence and individual patients' particular disease experiences.

Tacking is a maneuverability strategy deployed by sailors when faced with the difficult task of sailing directly into the wind. When tacking, sailors use head-on winds to propel the vessel forward by turning the bow back and forth into the wind. I argue that medical professionals engage in a similar act when faced with the challenge of navigating uncertainty and patients' corporeal contingencies. To explain such a complex practice within deliberative contexts, I introduce the construct, **enthymematic elasticity**. By mobilizing topological theories that characterize communities of practice in terms of how they navigate change,[15] I argue that enthymematic elasticity is a topological tactic for mitigating uncertainty, making decisions, and indoctrinating new members into a community of practice. Findings have implications not just for the ways enthymematic elasticity might be modeled to accommodate other kinds of evidences in medical decisions; I also hope to contribute to ongoing conversations in the humanities about how to collaboratively navigate uncertainty and contingency, particularly when human lives are at stake.

The Enthymematic Premise: Probability or Plausibility?

There are at least two dominant schools of thought on the definition and nature of an enthymeme. On the one hand, enthymemes are characterized as incomplete syllogisms. For example, Dyck defines an enthymeme as "a syllogism in which one or more premises may be probable and a *topos* replaces implication."[16] Specifically, Dyck (drawing on Burnyeat) argues that *topoi*, or rhetorical common places, make up one of an enthymeme's premises.[17] What is important about this characterization is that enthymematic premises are embodied by "if-then" statements[18] made possible through binary relationships (e.g., Aristotle's *topos*, more/less, cause/effect).[19] Similarly, Scott, citing Jeffrey Walker, references this definition for an enthymeme: "a body of persuasion that presents a claim, foregrounds a stance, and motivates identification with this stance by invoking a chain of premises and a cluster of value-charged proofs."[20] The truncated syllogism hypothesis seems to imply that enthymemes are a formulaic approach to argumentation and reasoning. In other words, under the rubric of enthymeme-as-truncated-syllogism, enthymematic argumentation is a form of logical reasoning expressed as inflexible arguments about how to think, act, or proceed with future action.

Contrary to the enthymeme-as-truncated-syllogism hypothesis is a markedly different characterization: enthymemes are an ongoing deliberation between interlocutors (or speaker/audience) in which the actual argument, itself, is invented through the very process of such negotiations (i.e., negotiations about the applicability of particular premises in actual, real world circumstances). Under the rubric of enthymeme-as-negotiation is Miller and Bee's description of enthymematic argumentation in terms of Aristotle's *phronesis*.[21] They argue that, like *phronesis*, enthymematic argumentation is concerned less with truth and more with "what is good or bad in the eyes of men."[22] In other words, enthymemes perform practical, situationally dependent, suasive labor.[23] Unlike scientific or logical reasoning, enthymemes are concerned more with negotiating "expediency or harmfulness."[24] Key to Miller and Bee's characterizations of enthymeme is the importance of audience participation. That which is finally agreed upon as expedient or harmful requires "the judgment of hearers,"[25] which always includes "feelings and emotions."[26]

Other rhetorical theorists similarly make the case for enthymematic argumentation as something very different from scientific or logical reasoning. Under the rubric of enthymeme-as-plausibility is Walton's argument[27] that enthymemes are not syllogisms with a missing premise; rather, they are arguments that posit the plausibility of particular futures.[28] For Walton, the success of enthymematic argumentation rests in more than formal or structural criteria; persuasion is fortified by informal criteria, or the power of plausible inferences. Walton makes what I regard as an important move to warn readers against collapsing plausibility with probability, since probability "has been taken over by the statisticians."[29] Unlike probability, plausibility denotes a presumption "about the way things can be generally expected to go in a kind of situation that would (presumably) be familiar to anyone."[30] He invokes "data bases" as a way of describing interlocutors' mutual "grasp of the setting of an everyday, familiar type of situation;" this shared understanding might also be understood as a script, or "a body of knowledge shared by language users concerning what typically happens in certain kinds of stereotypical situations."[31] Here we see traces of Miller and Bee's argument: Affect, and "empathy … the ability to put oneself into a familiar situation" are central to enthymematic argumentation.[32]

More recent theorizing about enthymematic argumentation includes Brock's suggestion that enthymemes are actually "a rhetorical algorithm, an incomplete syllogism whose implicit, unstated completion is realized only when an audience is persuaded to perform that completion—optimally in a manner aligned with a rhetoric's intent."[33] Drawing on a key construct in digital literacy studies, "procedural rhetoric,"[34] Brock makes the case that an enthymeme's premises are rules or steps necessary for moving audiences from one argumentative space to another. Although Brock seems to rely on an outdated definition for enthymeme (i.e., enthymeme-as-incomplete-syllogism) and suggests that enthymemes are probabilistic arguments (unlike Walton's desire to replace probability with plausibility), he does not see these characteristics as precluding flexibility and the ability to respond to local conditions. For Brock, probability is not necessarily synonymous with fixity or rigidity. Rather, an enthymeme's probabilistic character hinges on a series of conditional procedures. Importantly, such conditional procedures are not impervious to glitches. Drawing on Boyle,[35] Brock states that sometimes "the algorithm's rhetorical contingency, its potential for glitches or unintended actions, overpowers its probability of successful intent."[36] What Brock's definition of the enthymeme contributes to the following analyses is his understanding of algorithms as anything but "objectively logical," "context-free," or "nondeliberative."[37]

Brock's understanding of enthymeme moves toward what I understand as a third camp regarding definitions and characteristics of enthymematic argumentation. When making decisions in scenarios wherein "conclusive proof is not to be had," enthymematic argumentation can be used topologically.[38] That is, enthymemes can be a topological tactic or method for opening up possibilities, constructing available means, and inventing a way forward. Enthymeme-as-method is best represented by Scott's understanding of enthymeme as an "'architectonic rhetorical structure valuable in the invention process,' a heuristic for 'discovering the *best* reasons.'"[39] Similarly, Walker states that enthymematic argumentation "both invents and structures suasive discourse."[40] Here, enthymemes entail "not only a generative structure for an argument but also the argument itself."[41] Under the rubric of enthymeme-as-method, possibilities for future action are produced concomitantly with (or during) the manufacturing of the argument, itself. There are not rules, procedures, or footsteps to follow. Where an argument finally arrives is, in effect, fashioned from the very material-discursive phenomena that created the collaboratively constructed conditions for deliberation in the first place (see also Rickert's "ambient rhetoric"[42]).

Of course, among the many contributors who help to collaboratively condition possibilities for future action are human interlocutors. Although he does not refer to it as such, after concluding a study of enthymeme-as-method, Crick[43] emphasizes the audience's role in co-constructing scientific knowledge.[44] Scientific rhetoric is, for Crick, constitutive; he invokes the notion of "thought experiments" to describe enthymemes' constitutive nature. Thought experiments are "a halfway house between the actual and the possible."[45] Such a characterization does not just help to chip away at the wall that is assumed to exist between contexts of justification and contexts of discovery.[46] Thought experiments also reiterate how important cooperation between speaker and audience are when constructing argumentative premises in scientific contexts.[47] In particular, enthymemes "invite an audience to visualize a set of initial conditions and then imagine what might happen in response to those conditions."[48] Enthymematic argumentation as method is a distinctly different approach from using enthymemes as a way to prove "immediately verifiable empirical results."[49]

Building on the aforementioned theories—specifically, that enthymemes are akin to thought experiments that are architectonically productive—in what follows, I hope to demonstrate that decision-making in the biomedical backstage does not rely on rigid procedures for constructing probabilities. Rather, medical professionals in the biomedical backstage employ a topological tactic that allows for flexibility, versatility, and negotiation.

Joanie's Case

One backstage biomedical context in cancer care is the tumor board. Tumor board meetings include medical professionals from a range of specialties and expertise who deliberate about how best to proceed with caring for a patient whose cancer diagnosis is particularly complex for one reason or another. In Appendix A, I provide actual field notes from one patient's case presentation that, for the purposes of this project, I have named "Joanie."[50] The line numbers in Appendix A correlate with the following analyses, which include: (a) an analysis of the rhetorical moves made during Joanie's patient case presentation, and (b) an analysis of the evidential premises upon which each of those rhetorical moves are based.

Appendix A includes 131 lines from transcripts of tumor board deliberations about Joanie's case. Empty lines are used to delineate a shift in the deliberation, either in subject matter or speaker. With the exception of empty lines, I analyzed all lines for the rhetorical moves that speakers made while deliberating about Joanie's case. Table 11.1 provides a trace of the rhetorical moves made during Joanie's patient case presentation.

Table 11.1 Rhetorical moves made during Joanie's patient case presentation

Line #s	*Rhetorical moves*
1–3	The oncologist presents Joanie's age and relevant diagnostic details.
3–5	The oncologist describes how the profession would stage Joanie's cancer.
8	The oncologist discusses what current data suggests they should do in cases such as Joanie's.
9–10	The oncologist describes that Joanie declined treatment.
12–13	The oncologist describes that Joanie felt something was wrong.
15–17	The oncologist describes that she could not palpate Joanie's nodules.
22	The pathologist describes results from Joanie's CT scans.
23	The pathologist says he hasn't seen any statistics that will help with navigating Joanie's case.
25–29	The radiologist describes results from Joanie's PET scan.
34–42	The pathologist describes Joanie's results.
46	The oncologist declares that there is "no good data" that helps them navigate cases like Joanie's.
47–48	The oncologist describes that Joanie was tested for certain genetic biomarkers so they could better understand the nature of her cancer.
52–54	The chairperson of the meeting describes what the standard of care is in cases such as Joanie's.
56–57	The chairperson provides historical background on the treatment of breast cancer.
58–60, 70	The chairperson describes the profession's standard of care in cases like Joanie's.
72–74	The oncologist describes past experiences with such cancers.
76–78	The chairperson counters the relevancy of the oncologist's past professional experience by stating that Joanie's case is very different.
80–82	The oncologist describes that Joanie declined to participate in a clinical trial.
84–87	The chairperson provides details on how that clinical trial was run.
91	The chairperson describes what he knows he is and is not capable of in cases like Joanie's.
97–99	The chairperson provides greater detail on the standard of care in cases like Joanie's.
103–107	The chairperson provides definitions for certain criteria regarding the standard of care in cases like Joanie's.
111	The chairperson describes what the standard of care characterizes as a poor prognosis.
116–117	The chairperson describes additional details about how many rounds of chemotherapy the profession sees as standard in cases like Joanie's.
119–124	The oncologist and chairperson describe their experience treating a mutual patient with chemotherapy.
126–128	The oncologist describes how she has manipulated the treatment protocol stated by the standard of care so that it is more effective for a patient.
130–131	The chairperson describes experience with conducting the kind of manipulations described by the oncologist in ways that will not damage the patient's lungs and heart.

Dr. Thomas (oncologist): A 54-year-old patient presents with recurrence of cancer...original cancer was on the right side. Dr. Thomas declares that she "doesn't have her chart," but "my recollection was..." it was a diffuse tumor which means you stage the cancer by the largest palpable area. She was one node positive, so they gave her chemo. She was also hormone positive, so they take her to the maximum. (These are all events leading up to her later meeting with Dr. Thomas.)

Fig. 11.1 Snapshot of lines 1–6 from Joanie's patient case presentation

Line numbers from tumor board transcripts in Appendix A align with the second column labeled "Rhetorical Moves" in Table 11.1. To illustrate, observe the snapshot of lines 1–6 in Fig. 11.1.

The six lines from transcripts in Fig. 11.1 correlate with the following rhetorical moves:

(lines 1–3) The oncologist presents Joanie's age and relevant diagnostic details.
(lines 4–5) The oncologist describes how the profession would stage Joanie's cancer.

In addition to distilling rhetorical moves during Joanie's patient case presentation, I also associate those rhetorical moves with particular enthymematic premises (see Table 11.2).

In Table 11.2, line numbers from transcripts are indicated in the far left column and associated rhetorical moves are described in the far right column. The middle column provides readers with the enthymematic premises upon which each of those rhetorical moves is associated.

Specifically, I identify the following five enthymematic premises at work during deliberations about Joanie's case:

- Patient particulars (references to details about Joanie's unique disease and concomitant corporeal experiences)
- Standard of care (references to the National Comprehensive Cancer Network's official standard for how to care for a patient, given certain test results and disease experiences)
- Previously published studies (references to results [e.g., statistics, data, study design] from clinical trials)
- Past practice (references to a clinician's understanding of past, generalizable experiences within the cancer care community)
- Professional experience (references to a specific clinician's unique experience in cancer care).

Table 11.2 Enthymematic premises and rhetorical moves during Joanie's patient case presentation

Line #s	*Enthymematic premises*	*Rhetorical moves*
1–3	Patient particulars	The oncologist presents Joanie's age and relevant diagnostic details.
3–5	Standard of care	The oncologist describes how the profession would stage Joanie's cancer.
8	Previously published studies	The oncologist discusses what current data suggests they should do in cases such as Joanie's.
9–10	Patient particulars	The oncologist describes that Joanie declined treatment.
12–13	Patient particulars	The oncologist describes that Joanie felt something was wrong.
15–17	Patient particulars	The oncologist describes that she could not palpate Joanie's nodules.
22	Patient particulars via medical images	The pathologist describes results from Joanie's CT scans.
23	Previously published studies	The pathologist says he hasn't seen any statistics that will help with navigating Joanie's case.
25–29	Patient particulars via medical images (radiological)	The radiologist describes results from Joanie's PET scan.
34–42	Patient particulars via medical images (pathological)	The pathologist describes Joanie's results.
46	Previously published studies	The oncologist declares that there is "no good data" that helps them navigate cases like Joanie's.
47–48	Patient particulars	The oncologist describes that Joanie was tested for certain genetic biomarkers so they could better understand the nature of her cancer.
52–54	Standard of care	The chairperson of the meeting describes what the standard of care is in cases such as Joanie's.
56–57	Past practice	The chairperson provides historical background on the treatment of breast cancer.
58–60, 70	Standard of care	The chairperson describes the profession's standard of care in cases like Joanie's.
72–74	Professional experience	The oncologist describes past experiences with such cancers.
76–78	Patient particulars	The chairperson counters the relevancy of the oncologist's past professional experience by stating that Joanie's case is very different.
80–82	Patient particulars	The oncologist describes that Joanie declined to participate in a clinical trial.

(*continued*)

Table 11.2 (continued)

Line #s	*Enthymematic premises*	*Rhetorical moves*
84–87	Previously published studies	The chairperson provides details on how that clinical trial was run.
91	Professional experience	The chairperson describes what he knows he is and is not capable of in cases like Joanie's.
97–99	Standard of care	The chairperson provides greater detail on the standard of care in cases like Joanie's.
103–107	Standard of care	The chairperson provides definitions for certain criteria regarding the standard of care in cases like Joanie's.
111	Standard of care	The chairperson describes what the standard of care characterizes as a poor prognosis.
116–117	Standard of care	The chairperson describes additional details about how many rounds of chemotherapy the profession sees as standard in cases like Joanie's.
119–124	Professional experience	The oncologist and chairperson describe their experience treating a mutual patient with chemotherapy.
126–128	Past practice	The oncologist describes how she has manipulated the treatment protocol stated by the standard of care so that it is more effective for a patient.
130–131	Past practice	The chairperson describes experience with conducting the kind of manipulations described by the oncologist in ways that will not damage the patient's lungs and heart.

Said simply, cancer care professionals tack back and forth between one or more of these five enthymematic premises, or *topoi*, when navigating the uncertainty posed by Joanie's case. In transcript lines 1–6 (Fig. 11.1), for example, medical professionals rely on references to Joanie's unique patient details (age, cancer site) and references to the cancer community's standards of care ("which means you stage the cancer by … ").

During Joanie's patient case presentation, medical professionals make rhetorical moves that rely on the suasiveness of evidential premises that are at once specific to Joanie's unique case and also generalizable to the practice of cancer care. See how in Table 11.2, for example, deliberators move seamlessly between discussion of Joanie's CT scan results in line 22 to, in the next line, generalizable evidence vis-à-vis previously published studies. The same rhetorical tacking technique occurs a few lines later when **first** Joanie's pathology results are described in lines 34–42, **second** generalizable evidences are seamlessly integrated in line 46, and **third** deliberators return back to Joanie's individual and unique genetic biomarkers in lines 47–48. On its own, a single enthymematic premise—whether it is the suasiveness of previously published studies invoked in Appendix A, lines 84–87, a standard of care invoked moments later in line 97, or a reference in line 91 to one Dr. Neely's professional experience (which serves to link the other two evidential premises)—does not carry enough rhetorical weight to tip the tumor board's deliberative balance in one direction or another. Rather, what appears to occur during discussion about Joanie's case is that deliberators tack back and forth between enthymematic premises; along the way, tumor board members implicitly assess the value of each argumentative move based on how each enthymematic premise relates to one another.

It is impossible, therefore, to discern which isolated rhetorical moves and concomitant evidential premises are considered more or less suasive. Collectively, rhetorical moves propel the deliberation forward as a kind of argumentatively powerful evidential amalgamation. One might say that enthymematic premises are imbued with argumentative power only when considered in relationship to other enthymematic premises. For example, the professional experience that Dr. Neely invokes in Appendix A, line 91 is made more meaningful within the context of Joanie's case when referenced alongside studies and statistics. Medical decision-making is a matter of fit, therefore. Deliberators tack back and forth between evidential premises in order to build a collective sense about what is the best fit for treating Joanie's complex cancer diagnosis.

Required for such an approach to argumentation or deliberative decision-making, then, is a commitment to rhetorical flexibility. Such rhetorical flexibility affords the inventive space necessary for deliberators to engage in thought experiments.[51] Rather than making decisions based on probabilities,[52] deliberators in Joanie's case rely on the argumentative plausibility ultimately generated by deliberators' commitment to enthymematic elasticity.

Tacking Back and Forth as Technoscientific Co-Production

Much of the technoscientific labor that occurs in the biomedical backstage is characterized as "co-production."[53] By analyzing rhetorical moves and their underlying enthymematic premises, I argue that medical evidences continue the process of co-production once they enter spheres of deliberation.

In other words, enthymematic argumentation helps to manufacture evidential meaningfulness. It may be said, therefore, that enthymematic elasticity is one tactic for navigating and negotiating medical uncertainty amid corporeal contingency.

Findings from an analysis of Joanie's case points to the nuanced ways in which medical practitioners in this setting demonstrate what Bowker and Star (borrowing from Suchman and Trigg[54]) call "artful integration."[55] That is, tumor board deliberators find ways of tacking back and forth between multiple (and sometimes competing) enthymematic premises during their deliberation. For example, while it would seem in Appendix A, lines 97–117 that the National Comprehensive Cancer Network's official standard for how to manage Joanie's diagnosis should outweigh all other enthymematic premises, medical professionals do not see the standard of care as the default directive for how to move forward. Otherwise, referencing the standard of care in Joanie's case may have halted deliberations. Rather, the standard of care as an enthymematic premise appears to propel deliberations forward into new territory—in particular, the shared experience between Dr. Neely and Dr. Thomas about their mutual patient with recurrent breast cancer. The standard of care as an enthymematic premise propels deliberators into a discussion about a radically *non*standard patient and subsequent possibilities for treatment (Appendix A, lines 123–126). Cancer care deliberators tack back and forth between evidential premises as a way to advance, argumentatively.

I do not have access to what medical professionals officially presented to Joanie as her options for moving forward with her care, but tumor board meeting transcripts suggest that they found a way to integrate the standard of care into other evidential premises—including past practice, professional experience, previously published studies, and Joanie's unique disease experiences. In fact, it appears as if in lines 113–131, deliberators are, in a way, attempting to hack the standard of care as an evidential premise. While it is clear that Dr. Neely and Dr. Thomas are well versed in the standard of care (i.e., you only give one round of radiation therapy), by referencing their professional experiences with a mutual patient, they point to the pliability of this particular enthymematic premise: "There are some 'manipulations' we can do" (Appendix A, line 126). Dr. Neely and Dr. Thomas are aware that "different situations call for different measures" (Appendix A, lines 116–117). A range of solutions (or "manipulations") is made possible when deliberators tack back and forth between evidential premises, allowing such a practice to propel them forward in as of yet undiscovered ways. Enthymematic elasticity enables rhetorical invention.

Around the table during tumor board deliberations are those who have or will come into contact with Joanie (i.e., the pathologist, radiologist, oncologist, and tumor board chairperson). These persons are the primary providers of the deliberation's evidential premises. But there are others in the room as well, including medical students, nurses, case managers, and other hospital personnel who may never have contact with Joanie. These attendees come to hear Joanie's case for its educational value. In fact, many receive continued medical education credit for attending tumor board meetings. In addition to providing a backstage space for deliberation, tumor board meetings are also an education space, therefore—a place wherein those in attendance can become indoctrinated into the backstage deliberative procedures characteristic of this particular community of practice. From this perspective, new members learn how to manage "the tension between naturalized categories on the one hand and the degree of openness to immigration on the other."[56] New members learn how to engage in enthymematic elasticity. They learn what it looks like to value plausibility over probability. Rather than linear or causal approaches to diagnosis, prognosis, and treatment, tumor board deliberators model topological approaches for determining the best interventional fit for each individual patient.

Medical professionals invoke enthymematic premises that are "both plastic enough to adapt to local needs and constraints of the several parties employing them" while also "robust enough to maintain a common identity across sites."[57] Akin to Bowker and Star's notion of a boundary infrastructure, a tumor board's enthymematic elasticity demonstrates how to have both "sufficient play to allow for local variation" as well as a "sufficient consistent structure to allow for the full array of bureaucratic tools (forms, statistics, and so forth) to be applied."[58] Importantly, deliberations that employ a commitment to enthymematic elasticity pose the potential to provide "a non-ego-based, number-free and metric-free account of experience that respects evidence of continuous lived experience but does not reduce to sense perception or ego-centered experience."[59] That is, such a topological tactic is an "alternative" to or "thickening of" traditional, "quantitative methods" (242). Said another way: enthymematic elasticity "radically de-center[s]" and "de-anthropomorphize[s]" the experience of disease and all subsequent decision-making about it.[60]

Rather than offering a critique of medical practice as an inflexible, dictatorial enterprise, this chapter points to the productive possibility of integrating what already happens in the biomedical backstage—that is, a (seemingly ad hoc) commitment to enthymematic elasticity—into what could be a rhetorically rich medical education curriculum. In a very real and practical sense, evidential premises' pliability opens up diagnostic and prognostic possibilities rather than closing them down. Given the ways in which a commitment to enthymematic elasticity is a valued practice, future scholars might partner with medical professionals to propose a model for how to integrate a wide range of evidences (especially those that resist quantification—for example, patient-reported outcomes) into medical deliberations.

To illustrate the ephemerality of embodied evidence, at the beginning of this chapter I invoked Sha's smoke metaphor. Recall from Sha's smoke metaphor that the assumedly rational or empirical approach to determining its location was to ask each person to raise their hand once they could smell the smoke. And yet, Sha argues: "different people have different sensibilities and each person may be more or less sensitive to smoke."[61] There are, therefore, limitations to perspectival techniques for locating and attending to in flux phenomena. Like Joanie's cancer, there are limitations to understanding the smoke only in ways that are either causal or experiential. There are limitations to understanding the smoke without attending to what Sha calls an "amalgam"—a kind of evidential admixture "of the physical particles in motion, the people's physiologies, and the phenomenological expectation set by the asking."[62] Topological approaches to medical decision-making like tacking back and forth between varying and often times competing evidential premises are one way of building such evidential amalgamations. They are one way of doing evidence-based medicine without having to exchange the argumentatively inventive power of patients' unique disease experiences for (assumedly more) objective, generalizable evidences embodied by statistical frequencies and standards of care. Unlike linear or causal techniques, navigating medical uncertainty amid corporeal contingencies by practicing enthymematic elasticity does not foreclose evidential variation and argumentative flexibility. A commitment to enthymematic elasticity in the biomedical backstage is one step toward ensuring that medical decision-making honors both human fragility and evidential complexity.

Appendix A: Joanie's Patient Case Presentation

Dr. Thomas (oncologist): A 54-year-old patient presents with recurrence of cancer…original cancer was on the right side. Dr. Thomas declares that she "doesn't have her chart," but "my recollection was …" it was a diffuse tumor which means you stage the cancer by the largest palpable area. She was one node positive, so they gave her chemo. She was also hormone positive, so they take her to the maximum. (These are all events leading up to her later meeting with Dr. Thomas.)

Dr. Thomas says that "the data suggests" to add aromatase inhibitor as a preventative treatment (she suggests this to the patient), and the patient declines. Dr. Thomas says that after she declined, "we were just following along…"

Dr. Thomas says she isn't sure who saw the patient first, either her or the surgeon but the patient feels that something is wrong.

Dr. Thomas says that during the examination she could not palpate anything, could not feel anything, could not feel any skin nodules. Dr. Thomas says that when in doubt she "always believes the patient." So the patient goes to see the surgeon.

[Lights dim, attention is given to front screen where MRI of breast is shown, a red pointer is used throughout the presentation.]

The radiologist describes the CT, refers to the cancer as "a thing," describes stats and values and says he hasn't "seen" any stats.

He refers to one spot on the image as the "area of palpable concern." He also mentions that there may be one other thing that is notable to mention that may not have been on the original report; shows an image of the whole body that appears grayish and whitish and says that whoever "read" the PET scan thought that "this area" [points with pointer] must have been a scar. But radiologist advises that thay "may want to check into this."

[Attention is given to front screen still where images of cells are shown and a red pointer is used throughout the presentation]

The pathologist describes the biopsied cells as being "scattered cells;" [the designer of the image used arrows to designate where the cancer had "infiltrated the lobular…"]. The pathologist then goes on to describe that the cells have a "scattered, buckshot appearance" easily missed at "lower power." He then goes on to describe that the nuclei are "positive," again uses the term "infiltrate," describes the blue/grey counterstain. He goes on to describe the "lower power view." Also describes the dermis location, then the subcutaneous location (or beneath the surface). Pathologist refers to specific areas as "areas of interest," again mentions their "scattered, buckshot appearance," and describes how the cells have "lined up in a row," or what he calls "Indian filing."

[Lights come back on]

Dr. Thomas declares that there is "no good data" on this kind of situation. She goes on to describe that "tumors behave differently." She explains that she tested for HER II and the FISH Study.

[Question from audience about how best to treat the patient.]

[Answer from Dr. Neely:] We are to resect the disease, get clear margins; then we have a good prognosis. That's how to get her disease free. This is what you do when disease comes back locally and you are after "local control."

Dr. Neely goes on to describe the long history of how breast cancer has been treated. In particular, there was more of an emphasis on breast conservation. We try to use "categories that pigeonhole which patients will do badly, or predict good or not so good outcomes." Our "recommendation" for locally recurrent breast cancer is to "manage the problem as if it were the original problem. . .one recurrence predicts anthor."

Dr. Thomas adds that "we did offer her a study, but you have to have clear margins in order to participate in the study and no metastatic disease based on the "imaging" and then you have to agree to either continued observation after the radiation or chemotherapy."

[Question from audience]: Do you have to go through the ribs, do you remove the chest wall?

[Answer from Dr. Neely]: No, we don't "dip into it, we use it as a boundary."

Dr. Thomas adds that "we've all seen tumors that wrap themselves around the chest wall and literally strangle the woman. . .they infiltrate. . ." [lots of gestures here]; "the tumor goes beyond the margins."

Dr. Neely adds that this is an entirely different case. Thetumor is in and around the scar; there are nodules, it's not deep, it is subcutaneous [makes small circle motion]. He says this is a "radial concern," not "depth." He says,,"that's an entirely different entity."

Dr. Thomas says the patient declined the study even though she was eligible. She did not get chemo. She is still hormone sensitive, though, so that was the only "non-toxic option or thing to do."

Dr. Neely refers back to a study; says they "couldn't get enough women to agree to radiation after removal of the breast." He says there is all that "brewing literature out there that has never been proven that radiation after the mastectomy helps." He makes reference to the quadrants of the tumor (space is used as a predictive device).

Someone from audience questions Dr. Neely about the efficacy of this approach.

Dr. Neely says, "I cannot do what surgeons can do—local control cannot happen with only radiation."

[Question from the audience:] But other cancers (lung, etc.) can be treated with radiation and can be cured.

Dr. Neely: Well, yes but I specifically said breast cancer. There are other diseases, yes, but not breast cancer. The paradigm is that we "presume microscopic residual" disease and "must get a clear margin."

[Question from audience]: "Well what is a clear margin?"

Dr. Neely: NSAPB has two definitions; one is "no cancer on the ink surface." However as is the case with most very "subjective parameters," there are groups that are much more conservative and consider 5 mm clear, some a 1 mm margin of normal tissue. The key is to look at the space in between the high side and the low side, look at the in between.

[Question from audience]: "How about dermal invasion?"

Dr. Neely: Yeah once it gets into the skin that's bad, T4, worst prognosis.

[Question from audience]: Can you re-radiate someone who has previously had breast cancer, has had a mast and already had radiation?

Dr. Neely: The standard is one round—you don't keep re-treating, obviously different situations call for different measures.

Dr. Neely and Dr. Thomas discuss an example of a patient they share who has really bad breast cancer recurrence and they've continued to retreat her with radiation therapy. They describe her chest wall as now being "rock hard, woody, and unexaminable." They describe how they just hold their breath thinking that she is going to come in with something else. Dr. Neely describes that she must have received "well over 20,000 RADS" at this point.

Dr. Thomas: There are some "manipulations" we can do though. For instance,give patients chemo so that their skin is more sensitive to the radiation so the dose doesn't have to be as high.

Dr. Neely: with CT scans and computer-based RAD we can shield the lungs and the heart pretty well now when treating the chest wall.

Notes

1. Judith Butler, *Frames of War* (New York: Verso Books, 2009), 35.
2. Christa Teston, *Bodies in Flux: Scientific Methods for Negotiating Medical Uncertainty* (Chicago: University of Chicago Press, 2017).
3. Christa Teston, "A Grounded Investigation of Genred Guidelines in Cancer Care Deliberations," *Written Communication* 26 (2009), 320–348.
4. Christa Teston, "Moving From Artifact to Action: A Grounded Investigation of Visual Displays of Evidence During Medical Deliberations," *Technical Communication Quarterly* 21 (2012), 187–209.
5. Teston, *Bodies in Flux.*
6. My use of "backstage" derives from Erving Goffman's investigation into differences between communication in the frontstage. Unlike fronstage communication, backstage communication is limited to insiders. According to Ellen Barton, backstages in medical contexts are limited to those who "coconstruct" medical discourse and decisions (71).
7. Xin Wei Sha, "Topology and Morphogenesis," *Theory, Culture & Society* 29 (2012), 227.
8. Eleni Ikoniadou, *The Rhythmic Event: Art, Media, and the Sonic* (Cambridge: The MIT Press, 2014), 41.
9. Ibid., 42 (italics in the original).
10. See also Phil Bratta, "Rhetoric and Event: The Embodiment of Lived Events," *Enculturation: A Journal of Rhetoric, Writing, and Culture* (2015).
11. I acknowledge the study's limitations since analyses thus far are limited to Western, allopathic medicine.
12. Ed Dyck, "Topos and Enthymeme," *Rhetorica: A Journal of the History of Rhetoric* 20 (2002), 105–117.
13. J. Blake Scott, "The Public Policy Debate over Newborn HIV Testing: A Case Study of the Knowledge Enthymeme," *Rhetoric Society of America* 32 (2002), 57–83.
14. Susan Leigh Star, "Topology and Morphogenesis," *Theory, Culture & Society* 29 (2012), 601.
15. See Celia Lury, Luciana, Parisi, and Tiziana Terranova, "Introduction: The Becoming of Topological Culture," *Theory, Culture & Society* 29 (2012), 3–35.
16. Ed Dyck, "Topos and Enthymeme," *Rhetorica: A Journal of the History of Rhetoric* 20 (2002), 111.
17. Dyck adds nuance to this argument by reminding readers that Aristotle's original proposal was that premises may be presented in one of three ways—as a demonstrative syllogism, which is the case if the premises "are true or accepted as true" (110); as a dialectic, which is the case if premises are "chosen for the sake of argument or are reputable opinions" (110); or

rhetorically, which is the case if the premises are also dialectical and "its subject is that with which the divisions of rhetoric (deliberative, forensic, and epideictic oratory) are concerned" (110).

18. Ibid., 109.
19. Scott makes a similar observation about the prevalence of such a characterization in rhetorical scholarship. In an analysis of public policy debates about compulsory HIV testing for newborns, he opens by reminding readers of enthymemes' traditional definition: a "truncated syllogism based on probable rather than certain premises" (57).
20. Scott, "The Public Policy Debate," 57.
21. Arthur B. Miller and John D. Bee, "Enthymemes: Body and Soul," *Philosophy and Rhetoric* 5 (1972), 201–214.
22. Ibid., 205.
23. For Miller and Bee (ibid.), "feelings and emotions," or what they call "the affective component" is "inherent in the enthymeme," and as such is "the essence of Aristotle's concept of the enthymeme as practical reasoning" (201). Returning to its etymological roots, in particular, *thymos*, Miller and Bee note that Aristotle's enthymeme originally invoked the heart or soul and could be equated with such notions as to "take to heart, be concerned or angry at, … form a plan, … infer, conclude…." (202). Said simply: "for action to occur, there must be appetite" (203).
24. Ibid., 212.
25. Ibid., 212.
26. Ibid., 213.
27. Douglas Walter, "Enthymemes, Common Knowledge, and Plausible Inference," *Philosophy and Rhetoric* 34 (2001), 93–112.
28. Walton states that it is a "'misapprehension' of his [Aristotle's] meaning to think that he conceived of an enthymeme as a syllogism in which one premise is suppressed" (98).
29. Ibid., 98.
30. Ibid., 101.
31. Ibid., 108.
32. Miller and Bee, "Enthymemes: Body and Soul," 104.
33. Kevin Brock, "Enthymeme as Rhetorical Algorithm," *Present Tense: A Journal of Rhetoric in Society* 4 (2014), 2.
34. See Ian Bogost, *Persuasive Games: The Expressive Power of Video Games* (Cambridge: The MIT Press, 2007) and Annette Vee, *Proceduracy: Computer Code Writing in the Continuum of Literacy* (Madison: University of Wisconsin Press, 2010).
35. Casey Boyle, "The Rhetorical Question Concerning Glitch," *Computers and Composition* 35 (2015), 12–29.

36. Brock, "Enthymeme as Rhetorical Algorithm," 4.
37. Ibid., 5.
38. Myles F. Burnyeat, "Enthymeme: Aristotle on the Rationality of Rhetoric," *Essays on Aristotle's Rhetoric* (1996), 99.
39. Scott, "The Public Policy Debate," 57.
40. Jeffrey Walker, "The Body of Persuasion: A Theory of the Enthymeme," *College English* 56 (1994), 62.
41. Scott, "The Public Policy Debate," 58.
42. Thomas Rickert, *Ambient Rhetoric: The Attunements of Rhetorical Being* (Pittsburgh: University of Pittsburgh Press, 2013).
43. Nathan Crick, "Conquering Our Imagination: Thought Experiments and Enthymemes in Scientific Argument," *Philosophy and Rhetoric* 37 (2004), 21–41.
44. Crick notes Jasinski's (2001) summary: "the sophistic view of an enthymeme" acknowledges its "essentially polysemic character" (qtd. in Crick 24).
45. Ibid., 40.
46. Sorenson (1992) makes a similar case: "a paragraph expressing a thought experiment is actually an enthymeme" (214). Sorenson even goes so far as to say that "the enthymematic ploy gives you plenty of room to maneuver" (214).
47. Bitzer (1991) makes a similar argument: "the successful building of arguments depends on cooperative interaction between the practitioner and his hearers" (407).
48. Crick, "Conquering Our Imagination," 22.
49. Ibid., 22.
50. This and all names used throughout are pseudonyms.
51. See Crick, "Conquering Our Imagination."
52. See Walton, "Enthymemes."
53. See, for example, Sheila Jasanoff, *States of Knowledge: The Co-Production of Science and the Social Order* (New York: Routledge, 2004).
54. Lucy A. Suchman and Randall H. Trigg, "Artificial Intelligence as Craftwork" (1993).
55. Geoffrey C. Bowker and Susan Leigh Star, *Sorting Things Out: Classification and its Consequences* (Cambridge: The MIT Press, 2000), 292.
56. Bowker and Star, *Sorting Things Out*, 295.
57. Ibid., 297.
58. Ibid., 314.
59. Sha, "Topology and Morphogenesis," 242.
60. Ibid., 243.
61. Ibid., 227.
62. Ibid., 227.

Bibliography

Barton, Ellen. 2004. "Discourse Methods and Critical Practice in Professional Communication: The Front-Stage and Back-Stage Discourse of Prognosis in Medicine." *Journal of Business and Technical Communication* 18: 67–111.

Bitzer, Lloyd. 1991. "Aristotle's Enthymeme Revisited." *Quarterly Journal of Speech* 45: 399–408.

Bogost, Ian. 2007. *Persuasive Games: The Expressive Power of Video Games.* Cambridge: The MIT Press.

Bowker, Geoffrey C., and Susan Leigh Star. 2000. *Sorting Things Out: Classification and its Consequences.* Cambridge: The MIT Press.

Boyle, Casey. 2015. "The Rhetorical Question Concerning Glitch." *Computers and Composition* 35: 12–29.

Bratta, Phil. 2015. "Rhetoric and Event: The Embodiment of Lived Events." *Enculturation: A Journal of Rhetoric, Writing, and Culture.* Accessed June 1, 2016. http://enculturation.net/rhetoric-and-event

Brock, Kevin. 2014. "Enthymeme as Rhetorical Algorithm." *Present Tense: A Journal of Rhetoric in Society* 4: 1–7.

Burnyeat, Myles F. 1996. "Enthymeme: Aristotle on the Rationality of Rhetoric." In *Essays on Aristotle's Rhetoric*, Edited by Amélie Oksenberg Rorty. Berkeley: University of California Press.

Butler, Judith. 2009. *Frames of War.* Brooklyn: Verso Books.

Crick, Nathan. 2004. "Conquering Our Imagination: Thought Experiments and Enthymemes in Scientific Argument." *Philosophy and Rhetoric* 37: 21–41.

Dyck, Ed. 2002. "Topos and Enthymeme." *Rhetorica: A Journal of the History of Rhetoric* 20: 105–117.

Goffman, Erving. 1978. *The Presentation of Self in Everyday Life.* London: Harmondsworth.

Ikoniadou, Eleni. 2014. *The Rhythmic Event: Art, Media, and the Sonic.* Cambridge: The MIT Press.

Jasanoff, Sheila. 2004. *States of Knowledge: The Co-Production of Science and the Social Order.* London: Routledge Press.

Lury, Celia, Luciana Parisi, and Tiziana Terranova. 2012. "Introduction: The Becoming Topological of Culture." *Theory, Culture & Society* 29: 3–35.

Miller, Arthur B., and John D. Bee. 1972. "Enthymemes: Body and Soul." *Philosophy and Rhetoric* 5: 201–214.

Rickert, Thomas. 2013. *Ambient Rhetoric: The Attunements of Rhetorical Being.* Pittsburgh: University of Pittsburgh Press.

Scott, J. Blake. 2002. "The Public Policy Debate Over Newborn HIV Testing: A Case Study of the Knowledge Enthymeme." *Rhetoric Society of America* 32: 57–83.

Sha, Xin Wei. 2012. "Topology and Morphogenesis." *Theory, Culture & Society* 29: 220–246.

Sorenson, Roy. 1992. *Thought Experiments*. Oxford: Oxford University Press.

Star, Susan Leigh. 2010. "This is Not a Boundary Object: Reflection on the Origin of a Concept." *Science, Technology, & Human Values* 35: 601–617.

Teston, Christa. 2009. "A Grounded Investigation of Genred Guidelines in Cancer Care Deliberations." *Written Communication* 26: 320–348.

———. 2012. "Moving from Artifact to Action: A Grounded Investigation of Visual Displays of Evidence During Medical Deliberations." *Technical Communication Quarterly* 21: 187–209.

———. 2017. *Bodies in Flux: Scientific Methods for Negotiating Medical Uncertainty*. Chicago: University of Chicago Press.

———. forthcoming. "Pathologizing Precarity." In *Precarious Rhetorics*, ed. Wendy S. Hesford, Adela C. Licona, and Christa Teston. Under review at The Ohio State University Press.

Vee, Annette. 2010. *Proceduracy: Computer Code Writing in the Continuum of Literacy*. Madison: University of Wisconsin Press.

Walker, Jeffrey. 1994. "The Body of Persuasion: A Theory of the Enthymeme." *College English* 56: 46–65.

Walton, Douglas. 2001. "Enthymemes, Common Knowledge, and Plausible Inference." *Philosophy and Rhetoric* 34: 93–112.

Glossary

Dialectic: In classical philosophy, the dialectic is a technique for the cooperative pursuit of truth in any issue. Two opposing points of view (thesis and antithesis) are put into play to test each other with the goal of determining where the truth lies—in one or both. The "one" result is most common in Socratic dialectic, via which Socrates's interlocutors are gradually made aware of an internal contradiction in their thesis, a moment known as the *elenchus*. The "both" result is the typical outcome of Hegelian dialectic (cf. *Science of Logic*), via which the truths residing in both thesis and antithesis are fused in a higher-order synthesis that resolves the apparent contradiction.

Eidos/eide: In Aristotle's *Rhetoric*, the special topics, ones that characterize a particular technical discourse (of physics, ethics, etc.) (I.2, 1358a2–35): for example, Conservation of Momentum is an *eidos* of physics discourse; physicists return to it over and over to make and justify arguments; meanwhile, lawyers or physicians are unlikely to use it.

Endoxon/endoxa: Aristotle's term for widely shared beliefs (*doxa*) that a rhetor can count on their audience to hold and confirm (cf. *Rhetoric* I.2; *Topics* I.105; VI.iv.15–30). *Endoxa* ground the major premises (*protaseis*) of syllogisms: for example, in the syllogism "All men are mortal; Socrates is a man; therefore, Socrates is mortal," the major premise, "All men are mortal," is founded on common, probabilistic belief rather than scientific

© The Author(s) 2017
L. Walsh, C. Boyle (eds.), *Topologies as Techniques for a Post-Critical Rhetoric*, DOI 10.1007/978-3-319-51268-6

demonstration. Accordingly, in modern argumentation theory, *endoxa* make a reappearance as the often-unspoken and unprovable "warrants" or "ideographs" of public argument.[1] For example, the argument, "A flat tax is best because it treats everyone equally" relies on the unstated premise that "Equality is best"; this argumentative warrant is also an ideograph in that it is widely supported among Americans even if very few might actually agree on a technical definition of equality.

Enthymeme: An Aristotelian concept articulating the basic unit of a probabilistic argument—a conclusion plus at least one reason for reaching it (*Rhetoric* I.2; II.22). While enthymemes are often referred to by rhetoricians and argument theorists as "incomplete syllogisms" (the conclusion and minor premise of a syllogism, missing the major premise [see *endoxon/endoxa* above]), in fact Aristotle never prescribes a particular formula for the enthymeme. The examples he gives sometimes include explicit statements of *endoxa* or major premises.

Idia: Also *idioi topoi*, *idiai protaseis*; see *eidos* above.

Kairos/kairoi: From the Greek for "season" or "opportunity." As a technical term, *kairos*, like *topos*, was originally borrowed from military contexts, the opening in an enemy's armor or defenses through which an attack could succeed. There is also a second craft derivation of the term from weaving that expressed the moment in which warp threads parted to allow the shuttle to pass the weft thread through.[2] Greek rhetoricians, both Sophistic and Socratic, put *kairos* into tension with *topos* in their theories: the successful rhetor would be one who had trained himself to perceive and seize the right positions (*topoi*) in order to benefit from fleeting and contingent opportunities (*kairoi*) to change his situation through skillful argumentation.

As every other original term in the rhetorical canon, *kairos* has undergone many reinterpretations in 2500 years of practice. These interpretations can be helpfully grouped under temporal and spatial emphases (*kairos* has both temporal and spatial connotations). The temporal interpretations of kairos are expressed by theories that emphasize an "exigence" that a successful rhetor must either generate or respond to in a timely fashion.[3] Spatial interpretations of kairos stress its structural features—the many human and non-human agents required to shape the opportunity or "opening."[4] Some more recent interpretations have attempted to blend temporal and spatial definitions of kairos. These approaches imagine an argument or discourse not as an analog flow but as a digital stream of micro-chasms that appear seamless only because they are bridged instinctually, by habitual routines. By this account they attempt to balance the agency of the rhetor with the agency of other actors in the rhetorical situation.[5]

Koinoi Topoi: Universally applicable *topoi* such as compare/contrast, cause and effect, division, definition, more and less, and so on. Aristotle's actual phraseology is "*topous de tous koinous homios panton*" (*Rhetoric* II.23). Cope coined the phrase *koinoi topoi* in 1867 as shorthand for the various ways in which Aristotle expressed the universal topics in the *Topics* and the *Rhetoric* II.19, 23–24.[6]

Locus/loci: The Roman version of *topos/topoi*, formulated in greatest detail by Cicero (*De Oratore* 2.162–172; *De Inventione*) and creatively reconciled with Aristotle's topical system by Boethius (*De Topicis Differentiis*). There are substantial technical differences among these systems, but these differences have been largely effaced with time: use of the Latin or Greek term among contemporary rhetoricians is due more to which system they have studied most closely—Cicero's or Aristotle's—than to technical disagreements about the nature of *loci* versus *topoi*. Perelman and Olbrechts-Tyteca preferred *locus/loci* in the *New Rhetoric*. The topical revival in technical communication studies preferred Aristotle's system.

Loci Communes: Literally "commonplaces." This was a Latin transliteration of Aristotle's "*topous de tous koinous homios panton*" (*Rhetoric* II.23). Perelman and Olbrechts-Tyteca use the Latin term technically to denote a system of their invention that synthesizes Aristotle's topics with his categories. Subsequent scholars have taken a different approach, employing the Latin and Greek versions of the term to express a subtle distinction. These scholars tend to use the term *topos/topoi* when they are talking about technical heuristics for inventing or structuring more complex arguments; by contrast, the term *locus* or commonplace is used to indicate "canned" or formulaic expressions of recurring arguments—as the commonplace "You do you" expresses a form of unengaged or dismissive disagreement.

Rhetoric: Refers to the ancient art of communication, in both the word's normal sense and the sense of community formation. While every society in the world has its own art of communication, whether explicit or implicit, the Greek system has been most developed and utilized in the West. The history of Greek rhetoric is commonly begun with the figure

Corax ("raven") of Syracuse, who set down some fragmentary principles of persuasive speech in the fifth century BC. Certainly, rhetorical principles circulated earlier; many are visible by inspecting speeches in Homer's epics. However, it was not until the fourth century that rhetorical treatises began to be preserved more or less intact. Socrates's student Aristotle assembled the best codified and most resilient theory of rhetorical practice in the *Rhetoric* and *Topics*. Meanwhile, his fellow student Plato produced important Socratic commentaries on rhetoric and communication (cf. *Phaedrus, Gorgias, Timaeus, Meno, Crito, Apologia*); and, speeches by private rhetoric instructors known as the Sophists (e.g., Gorgias, Protagoras, Antiphon, Thrasymachus) continued to circulate alternate theories of persuasion that focused more on non-rational than rational persuasion. Aristotle's contemporary Isocrates also produced rhetorical texts that hybridized Socratic and Sophistic theories (cf. *Antidosis*, *Against the Sophists*, *Panathenaicus*).

Roman rhetoricians studied and adapted these theories to practice in the Senate, Cicero (*On Oratory, On Invention*) and Quintilian (*Institutes of Oratory*) being chief among them. Hermogenes (*On Legal Issues*) codified the important doctrine of *stasis*, the habitual progress of questions in democratic deliberation from questions of fact and definition toward questions of value and finally of action (see entry below).

Augustine (*On Christian Doctrine*) helped translate and bowdlerize these pagan theories for the Christian era; Boethius (*Consolation of Philosophy*, *Topics*) attempted to standardize the proliferation of classical rhetorical systems for study by medieval students. But it was not until the Renaissance, when classical works were recuperated from Arabic translations, that rhetoric flourished again as an art. Important reinterpretations of classical principles for the purposes of education and self-formation were written by Erasmus, Ramus, and Vico. English rhetoricians such as Thomas Wilson and George Puttenham wrote tracts that adapted rhetoric for a rising bourgeois class in London.

With the Enlightenment, rhetoric took a back seat to dialectic, long favored by philosophers, who had been working gradually to divorce it from persuasive and political practices. In many ways, we find ourselves still living in the Enlightenment, particularly when politicians dismiss an argument as "just empty rhetoric," or Continental philosophers associate the art with propaganda and fascism.

However, in the Anglo-American tradition, rhetoric saw a revival beginning in the nineteenth century with teachers such as George

Campbell, Hugh Blair, and Richard Whatley, who sought to tie classical principles to emerging research on psychology and cognition for the purposes of reading and approaching audiences.[7] Their work was imported to the US and made the foundation of the "rhetoric and composition" programs that sprang up across land-grant institutions in the late nineteenth and early twentieth centuries. While the bread-and-butter of these departments was teaching farmers how to write essays in accordance with the land-grant commitment to a general education, these places provided fertile ground for continuing research into rhetorical theory and practice. John Dewey's pragmatism inspired many twentieth-century rhetorical scholars, including Richard McKeon, I. A. Richards, and Wayne Booth. Rhetoric developed subspecializations in history, literature, science, political economy, media studies, non-Western rhetorics, and public address, to name a handful of current emphases. Rhetorical theory in the twentieth century tended to find its motive force in cognate fields that had been developing rapidly while rhetorical research lay dormant in the eighteenth and nineteenth centuries: the legal philosopher Stephen Toulmin's argumentation theory was particularly well received, as was the dramatistic framework provided by Kenneth Burke in *Grammar of Motives* and *Rhetoric of Motives*; finally, a new major work from Europe, Perelman and Olbrechts-Tyteca's *New Rhetoric*, galvanized the field with its mixture of social psychology, pragmatic-linguistics, and classical rhetoric. Toward the end of the century, rhetoric took what is known as the "critical turn," as social constructivism, post-modernism, and Marxism informed new accounts of the political effects of persuasive speech.

Rhetoric finds itself now, as do many of the humanist disciplines, in a "post-critical" moment that seeks to heal the long Enlightenment detachment of intellectual from political practice. Rhetoricians are particularly attracted to works that blend or break down boundaries, such as those between the sciences and the humanities, those between verbal and other modes of expression, and those between academic institutions and communities. Throughout its 2500-year history, however, rhetoric has maintained its originary emphasis on *kairos* (see above), on the contingency of communication in time and space, an emphasis which sets it apart from other architectonic disciplines and makes it particularly well suited to intervene in the post-critical moment.

Stasis: Doctrine Systematized by Hermogenes of Tarsus (*On Legal Issues*) and developed by Quintilian and subsequent philosophers, *stasis* doctrine described the habitual sequence of *topoi* followed by legal inquiry in the Roman empire: questions of fact ("what happened?"), then questions of definition ("how do we classify this event?"), ending with questions of quality ("how good/bad is this classified event?"). The *stasis*, meaning "sticking point," was the question or point at which any given inquiry seemed to hang until enough data or consensus could be acquired to move forward. Four- and five-*stasis* systems have subsequently developed to model other deliberative communities, such as government panels and particular academic disciplines. However, all versions of *stasis* doctrine share these features: (1) an assumption that by default, inquiries move up-*stasis* over time from fact-gathering toward policy-making; (2) nevertheless, higher-*stasis* questions of value are involved from the outset, in determining which facts to gather; (3) therefore, on close examination, *stasis* "looping" occurs at almost every step of the inquiry, as the settling of complex questions requires many deliberative and fact-finding activities to be going on in tandem. In other words, the farther we stand back from *stasis* in any given inquiry, the more clear-cut the sequence appears, and the more closely we examine it, the more complex it appears.[8]

Topos/topoi: From the Greek for "place" or "stance," a rhetorical *topos* indicates a perspective from which a rhetor may invent an argument about a subject. While the term pre-existed Aristotle, he was the first to develop it into a full-fledged technical system of argumentation, in the *Topics* as well as Book II of the *Rhetoric*. In those works, Aristotle presented at least four different categories of *topoi*: the common, general, or universal, which could supposedly be used to generate arguments about any matter whatsoever (compare/contrast, cause and effect, definition, etc.); the so-called middle topics, which applied to rhetorical occasions concerning justice, praise or blame, and deliberation; the *endoxa*, which expressed values shared across the polity (liberty, fidelity, justice, etc.); and the *idia* or *eide*, which grew out of the organizing principles of specific technical discourses such as physics (e.g., conservation of momentum, the observer effect) or law (e.g., *habeas corpus*, legal standing, statutes of limitations, etc.).

After Aristotle and well into the Renaissance, rhetoricians continued to work seriously and programmatically with *topoi*, generating new systems of *loci communes*, often translated "commonplaces." Hermogenes, Cicero, Quintilian, Boethius, and Erasmus all contributed significantly to this effort. However, since these scholars worked at times from Aristotle's original catalog, at times from their predecessors' catalogs, and at times from their own intuitions and social milieus, the *topoi* proliferated, becoming contradictory and cumbersome.[9] In the late Medieval period, students might memorize and then reel off in the course of a dissertation more than 100 *topoi* in order to gain points with their professors. Furthering this "laundry list" abuse of topical systems, it was common practice for educated gentlemen in the Renaissance to keep "commonplace books," which were essentially litanies of witty quips or exotic facts with which they could salt their letters and impress their interlocutors.

Unsurprisingly, early modern rhetoricians such as Francis Bacon and George Puttenham rejected topical systems as rote formulae unworthy of true intellectuals.[10] Nor did Campbell, Whately, and the other nineteenth-century belles-lettres rhetoricians deem them useful hermeneutics for determining audience tastes. Serious engagement with topical systems of invention in Western rhetoric thus went on hiatus for over 350 years.

The recuperation of *topoi/loci* dates from the publication of *The New Rhetoric* in 1969, when Perelman and Olbrechts-Tyteca recast the *loci* as basic psychosocial principles and values that a speaker had to take account of in order to be persuasive. Thereafter, topical invention experienced a revival as part of cognitive programs in rhetoric and argument theory, with many scholars using *topoi* to describe the thought-styles of technical disciplines.[11] This revival was bolstered by a concurrent program in text linguistics that was searching for the logical connections that defined textual modes; variously called "discourse relations," "logical-semantic relations," or "coherence relations," these were essentially identical to *topoi* found in Aristotle's original catalog.[12]

The topical revival was derailed by the critical turn in the 1990s—specifically, by Gaonkar's critique of classical approaches to rhetoric of technoscience, the subfield of rhetoric that had most enthusiastically embraced topical methods.[13] Gaonkar argued that the classical *topoi* were heuristics for the production of texts and were never meant to function hermeneutically; further, since they developed out of specific practices in the ancient Athenian assembly, it was anachronistic and inappropriate to search for them in late-modern discourses of technoscience.

While classical scholars at the time offered cogent answers to Gaonkar's critique—namely, that the *topoi* had always been ambivalent with respect to the production and interpretation of discourse; that since classical agonistic habits of argument persisted in modern discourse, the socio-cognitive stances described by Aristotle's common *topoi* could also be expected to persist; and that the topical doctrine had never prescribed scholars a catalog from which to embark on a scavenger hunt, but rather the habit of inductively seeking out the controlling structures of the discourse at hand[14]—still critical scholars generally turned their backs on topical approaches.

However, at almost the same moment, spatial studies were entering the humanities from geography, bringing with them the notion of topology. Though these topologies were mathematical and philosophical in construction, it did not take long for them to catch the attention of rhetoricians still working with classical *topoi*—particularly the new-materialist rhetoricians. The old misreadings of *topoi* as rationalist, rigid, formulaic, strictly heuristic, and so on do persist.[15] Nevertheless, as evidenced by the contributions to this volume, a new generation of rhetorical scholars is actively engaged in recovering the pre-rational, non-linear, spatial-associative potential of *topoi* to model the habits of a discourse in ways that provide not only a hermeneutic of its power dynamics but also a heuristic for constructively intervening in it.

Warrant: In Toulmin argumentation, the often-unstated value or assumption that validates a basic argument or enthymeme (see above): for example, for the enthymeme "My opponent is unfit for office because he had an affair," the unstated warrant can be reconstructed as something like, "Only those who have been faithful in their marriages are fit for office." Warrants quite frequently reference *endoxa*, or communal values (see above).

Notes

1. Michael Calvin McGee, "The "Ideograph": A Link between Rhetoric and Ideology," *Quarterly Journal of Speech* 66, no. 1 (1980); S. E. Toulmin, *The Uses of Argument* (Cambridge, UK: Cambridge University Press, 1958).
2. Debra Hawhee, "Kairotic Encounters," in *Perspectives on Rhetorical Invention*, ed. J. Atwill and J.M. Lauer (Knoxville, TN: University of Tennessee Press, 2002); Thomas Rickert, "Location in the Wild: On Locating *Kairos* in Space-Time," in *The Locations of Composition*, ed. C.J. Keller and C.R. Weisser (Albany, NY: State University of New York Press, 2007).
3. Richard E Vatz, "The Myth of the Rhetorical Situation," *Philosophy & Rhetoric* 6, no. 3 (1973); J Blake Scott, "Kairos as Indeterminate Risk Management: The Pharmaceutical Industry's Response to Bioterrorism," *Quarterly Journal of Speech* 92, no. 2 (2006).
4. Lloyd F Bitzer, "The Rhetorical Situation," *Philosophy & Rhetoric* 1, no. 1 (1968).

5. T. Rickert, *Ambient Rhetoric: The Attunements of Rhetorical Being* (University of Pittsburgh Press, 2013), 77–107.
6. Sara Rubinelli, *Ars Topica: The Classical Technique of Constructing Arguments from Aristotle to Cicero* (Dordrecht, NL: Springer, 2009), 62.
7. James L Golden and Edward PJ Corbett, *The Rhetoric of Blair, Campbell, and Whately* (Carbondale, IL: SIU Press, 1968).
8. Jeanne Fahnestock and Marie Secor, "The Stases in Scientific and Literary Argument," *Written Communication* 5, no. 4 (1998); S Scott Graham, *The Politics of Pain Medicine: A Rhetorical-Ontological Inquiry* (University of Chicago Press, 2015), Chapter 6; Lynda Walsh, *Scientists as Prophets: A Rhetorical Genealogy* (New York: Oxford University Press, 2013), Chapter 5.
9. See Michael Leff, "Up from Theory: Or I Fought the Topoi and the Topoi Won," *Rhetoric Society Quarterly* 36, no. 2 (2006).
10. Carolyn S. Miller, "Aristotle's 'Special Topics' in Rhetorical Practice and Pedagogy," ibid. 17, no. 1 (1987).
11. Fahnestock and Secor, "The Stases in Scientific and Literary Argument"; Michael Leff, "The Topics of Argumentative Invention in Latin Rhetorical Theory from Cicero to Boethius," *Rhetorica* 1, no. 1 (1983); Carolyn S. Miller and Jack Selzer, "Special Topics of Argument in Engineering Reports," in *Writing in Nonacademic Settings*, ed. Lee Odell and Dixie Goswami (New York: Guilford Press, 1985); Lawrence J. Prelli, *A Rhetoric of Science: Inventing Scientific Discourse*, ed. Carroll C. Arnold, Studies in Rhetoric/Communication (Columbia, SC: University of South Carolina Press, 1989); Frans H. van Eemeren and Peter Houtlosser, "Strategic Manoeuvring in Argumentative Discourse," *Discourse Studies* 1, no. 4 (1999); Karl R. Wallace, "*Topoi* and the Problem of Invention," *Quarterly Journal of Speech* 58 (1972); W. Ross Winterowd, "'Topics' and Levels in the Composing Process," *College English* 34 (1973).
12. See for example John Swales, "Research Articles in English," in *Genre Analysis: English in Academic and Research Settings* (Cambridge, UK: Cambridge University Press, 1990); W. Mann and S. Thompson, "Relational Propositions in Discourse," *Discourse Processes* 9, no. 1 (1986); R. Longacre, *An Anatomy of Speech Notions* (Ghent, Belgium: Peter de Ridder Press, 1976); N. Asher and A. Lascarides, *Logics of Conversation* (Boston: Cambridge University Press, 2003); M. A. K. Halliday and R. Hasan, *Cohesion in English* (Boston: Longman, 1976).
13. Dilip Parameshwar Gaonkar, "The Idea of Rhetoric in the Rhetoric of Science," *Southern Journal of Communication* 58, no. 4 (1993).
14. See the collected responses in Alan G. Gross and William M. Keith, *Rhetorical Hermeneutics: Invention and Interpretation in the Age of Science* (Albany, NY: Suny Press, 1997).
15. Cf. Rickert, *Ambient Rhetoric: The Attunements of Rhetorical Being*, 60–70.

Bibliography

Asher, N., and A. Lascarides. 2003. *Logics of Conversation*. Boston: Cambridge University Press.

Bitzer, Lloyd F. 1968. "The Rhetorical Situation." *Philosophy & Rhetoric* 1, no. 1: 1–14.

Fahnestock, Jeanne, and Marie Secor. 1998. "The Stases in Scientific and Literary Argument." *Written Communication* 5, no. 4: 427–443.

Gaonkar, Dilip Parameshwar. 1993. "The Idea of Rhetoric in the Rhetoric of Science." *Southern Journal of Communication* 58, no. 4: 258–295.

Golden, James L., and Edward P.J. Corbett. 1968. *The Rhetoric of Blair, Campbell, and Whately*. Carbondale: SIU Press.

Graham, S. Scott. 2015. *The Politics of Pain Medicine: A Rhetorical-Ontological Inquiry*. Chicago: University of Chicago Press.

Gross, Alan G., and William M. Keith. 1997. *Rhetorical Hermeneutics: Invention and Interpretation in the Age of Science*. Albany: State University of New York Press.

Halliday, M.A.K., and R. Hasan. 1976. *Cohesion in English*. Boston: Longman.

Hawhee, Debra. 2002. "Kairotic Encounters." In *Perspectives on Rhetorical Invention*, ed. J. Atwill and J.M. Lauer, 16–35. Knoxville: University of Tennessee Press.

Leff, Michael. 1983. "The Topics of Argumentative Invention in Latin Rhetorical Theory from Cicero to Boethius." *Rhetorica* 1, no. 1: 23–44.

———. 2006. "Up from Theory: Or I Fought the *Topoi* and the *Topoi* Won." *Rhetoric Society Quarterly* 36, no. 2: 203–211.

Longacre, R. 1976. *An Anatomy of Speech Notions*. Ghent: Peter de Ridder Press.

Mann, W., and S. Thompson. 1986. "Relational Propositions in Discourse." *Discourse Processes* 9, no. 1: 57–90.

McGee, Michael Calvin. 1980. "The 'Ideograph': A Link Between Rhetoric and Ideology." *Quarterly Journal of Speech* 66, no. 1: 1–16.

© The Author(s) 2017

L. Walsh, C. Boyle (eds.), *Topologies as Techniques for a Post-Critical Rhetoric*, DOI 10.1007/978-3-319-51268-6

Miller, Carolyn S. 1987. "Aristotle's 'Special Topics' in Rhetorical Practice and Pedagogy." *Rhetoric Society Quarterly* 17, no. 1: 61–70.

Miller, Carolyn S., and Jack Selzer. 1985. "Special Topics of Argument in Engineering Reports." In *Writing in Nonacademic Settings*, ed. Lee Odell and Dixie Goswami, 309–341. New York: Guilford.

Prelli, Lawrence J. 1989. *A Rhetoric of Science: Inventing Scientific Discourse*, Studies in Rhetoric/Communication. Ed. Carroll C. Arnold. Columbia: University of South Carolina Press.

Rickert, Thomas. 2007. "Location in the Wild: On Locating *Kairos* in Space-Time." In *The Locations of Composition*, ed. C.J. Keller and C.R. Weisser, 61–90. Albany: State University of New York Press.

———. 2013. *Ambient Rhetoric: The Attunements of Rhetorical Being*. Pittsburgh: University of Pittsburgh Press.

Rubinelli, Sara. 2009. *Ars Topica: The Classical Technique of Constructing Arguments from Aristotle to Cicero*. Dordrecht: Springer.

Scott, J. Blake. 2006. "Kairos as Indeterminate Risk Management: The Pharmaceutical Industry's Response to Bioterrorism." *Quarterly Journal of Speech* 92, no. 2: 115–143.

Swales, John. 1990. "Research Articles in English." In *Genre Analysis: English in Academic and Research Settings*, 110–176. Cambridge: Cambridge University Press.

Toulmin, S.E. 1958. *The Uses of Argument*. Cambridge: Cambridge University Press.

Vatz, Richard E. 1973. "The Myth of the Rhetorical Situation." *Philosophy & Rhetoric* 6, no. 3: 154–161.

Wallace, Karl R. 1972. "*Topoi* and the Problem of Invention." *Quarterly Journal of Speech* 58: 387–395.

Walsh, Lynda. 2013. *Scientists as Prophets: A Rhetorical Genealogy*. New York: Oxford University Press.

Winterowd, W. Ross. 1973. "'Topics' and Levels in the Composing Process." *College English* 34: 707–708.

Index[1]

[1]Note: Page numbers followed by "n" denotes notes.

© The Author(s) 2017

L. Walsh, C. Boyle (eds.), *Topologies as Techniques for a Post-Critical Rhetoric*, DOI 10.1007/978-3-319-51268-6

The manufacturer's authorised representative in the EU is Springer Nature Customer Service Centre GmbH, Europaplatz 3, 69115 Heidelberg, Germany. If you have any concerns regarding our products, please contact ProductSafety@springernature.com

Printed and bound by CPI Group (UK) Ltd, Croydon, CR0 4YY

30/09/2026

02225649-0004